AF560284

IPR, Biosafety and Bioethics

NIPA® GENX ELECTRONIC RESOURCES & SOLUTIONS P. LTD.
New Delhi-110 034

About the Authors

Prof. (Dr.) Nirmal Mandal, a distinguished academic and researcher, currently serves as the Head of the Department of Agricultural Biotechnology, BCKV, Mohanpur. A doctorate from the illustrious Bose Institute, Kolkata, he has carved a niche in the field of in vitro culture technology, genetic transformation, and molecular biology. With a prolific academic career spanning over 28 years, he has mentored 28 Ph.D. scholars and authored numerous high-impact publications. Prof. Mandal imparts UG, M.Sc., and Ph.D. courses on Genome Engineering, Omics and Systems Biology, and Molecular Biology. A Fellow of the Society for Sciences, he remains a guiding force in cutting-edge agricultural biotechnology.

Prof. (Dr.) Sankar Kr Acharya, presently, **Dean, Post Graduate Studies;** former Head, Dept. of Agril Extension and Director, Extension Education, Bidhan Chandra KrishiViswavidyalaya, Mohanpur, WB, born on 6th October, 1960, started his career as Assistant Professor at BCKV in 1988 and has been in teaching, research and extension more than 36years. An erudite teacher as well as an elegant speaker, is internationally acclaimed for his unique research domain of Social Entropy and Energy Metabolism, Social Ecology and Environmental Sociology, Enterprise Ecology Framework, Conservation Stewardships: **Research Publication: 251 in National and International Journals. Book Publication: 119 books authored**. So far, he has delivered 49 keynote addresses and chaired 57 sessions in National and International congresses.

Expert Member, Agriculture Commission; Marketing and Extension Sub Committee, Govt. of West Bengal; Expert member WWF projects in BTR (Buxa Tiger Reserve Project); Expert member; DFID project on Primary Education (DPEP); Rapid Environment Impact Analysis.Co-PI of ICAR, NAHEP and World Bank funded project on Conservation Agriculture. So, far he has successfully guided 23 Ph D scholars and 97 M Sc scholars,. **He is the Fellow of ISEE, IARI, New Delhi, Fellow, BIOVED, Allahabad; Fellow, Society of community mobilization, IARI, New Delhi. He is now acting as editor, reviewer of several national and international journals** Awards and Distinctions: **Distinguish Scholar Award**, Krishisanskriti, New

Delhi; Certificate of Merit for standing First class first at M Sc(Ag) in Agricultural Extension, 1986; ;**Honorary Appointment to the Research Board of Advisors, The American Biographic Institute, USA** (2003) ;He has been honored to be selected as the Convener of Panel (PE-32) entitled the hunger, poverty and silence, of World Congress, IUAES, University of Manchester, UK, 2013 ;He has been honoured to be selected as the Convener of Panel (P-102) entitled ,' Uncertainties, Unpredictability and Marginalization: The impact and mitigation for survival of agriculture and humanity, of the 19th World Congress, IUAES-WAU, 14-20 October, 2023, New Delhi, India, 2023 ;Honoured to be nominated as Expert Member, Working group on Agricultural Extension, West Bengal State Agriculture Commission (2007) ;Honoured to be nominated by the Hon'ble Vice-Chancellor, BCKV to Act as an Expert Member, DPEP, A DFID, UK, project for planning and revamping primary education through POA of GOVT of India.(1995)Expert Member, Buxa Tiger Reserve Project funded by WWF(World Wide Wild Life Funding), 1998 ;**Dr Daulat Singh Memorial Award**, Society of Extension Education, **Dr. M.S. Swaminathan Award**, Brainware University and CRIJAF, ICAR; **Distinguished Professor and Academician Award**, Research Education Solution (RES), M S Swaminathan School of Agriculture, Centurion University(2022); **Eminent Scientist Award**, EIABT, Citwan Nepal. He is also the Expert Member, Research Council, Vidyasagar University, WB.

Prof Acharya is the **Achievers of German patent** (Utility model no.20202310273) :Novel LOT based inventory management system with radio controlled pallet racking for storage solution. He is the Fellow, West Bengal Academy of Science and Technology (WAST), 2024. He is also the fellow of ISEE, New Delhi, Fellow, BIOVED, Allahabad, Fellow, Mobilization, New Delhi. On academic missions, he visited **Italy, France, Germany, China, Sri Lanka, Vietnam, Bangladesh,** He has a rich profile of editorial and reviewers' contributions a different national and international Journals. Now, he has been invited to Zurich, Switzerland to deliver the Keynote address in International conference on Conservation Agriculture, June 22-23, 2025.

IPR, Biosafety and Bioethics

Concepts, Regulations and Applications in Biotechnology

N. Mandal
S. Acharya

NIPA® GENX ELECTRONIC RESOURCES & SOLUTIONS P. LTD.
New Delhi-110 034

NIPA® GENX ELECTRONIC RESOURCES & SOLUTIONS P. LTD.

101,103, Vikas Surya Plaza, CU Block
L.S.C. Market, Pitam Pura, New Delhi-110 034
Ph : +91-11-43860225, Mob.: +91 9717133558, 9540816132
E-mail: newindiapublishingagency@gmail.com
Website: www.nipaersources.com

Print ISBN: 978-93-58875-04-1
ebook ISBN: 978-93-58873-90-0

Composed and Designed by NIPA®.

Dedication

to the torchbearers of knowledge —
the students, researchers, and teachers —
whose pursuit of truth, ethics, and
innovation shall shape a safer and more
just biotechnological future

Preface

The ever-expanding frontier of biotechnology has brought forth not only remarkable scientific advancements but also a complex matrix of ethical, legal, and societal considerations. In this era of genomics, synthetic biology, genetic engineering, and bioprospecting, it has become imperative for the academic and research community to possess not only scientific expertise but also a firm grounding in the ethical, regulatory, and legal frameworks that govern biotechnological applications.

This textbook, has been meticulously prepared in alignment with the **revised Postgraduate Curriculum prescribed by the Indian Council of Agricultural Research (ICAR)** for the discipline of **Molecular Biology and Biotechnology**. It also holds substantial relevance for postgraduate students across general **Biotechnology**, **Agricultural Biotechnology**, **Life Sciences**, and **Interdisciplinary Biosciences** programs.

The book is structured to provide a comprehensive and pedagogically sound treatment of key areas including **biosafety regulations**, **intellectual property rights**, **bioethics**, **biopiracy**, and **benefit-sharing**. It systematically navigates through the scientific principles, national and international regulatory frameworks, legal instruments, ethical paradigms, and societal implications associated with biotechnological innovation and its deployment.

In crafting this volume, deliberate efforts have been made to integrate **updated statutory guidelines**, **real-world case studies**, **operational procedures**, and **critical reflections** on policy issues. Each chapter reflects a synthesis of current academic discourse, policy documentation, and regulatory practices, articulated in a manner that is **intellectually rigorous**, **conceptually lucid**, and **accessible to postgraduate learners** and early-career researchers.

Key features of this book include:

- **Syllabus-oriented structure**, facilitating easy adoption in university coursework.
- Emphasis on **Indian Acts, international conventions, and institutional frameworks**.

- Inclusion of **model forms**, **case studies**, and **appendices** for practical understanding.
- A robust **glossary** and **abbreviations section** to aid terminological clarity.

This textbook aspires to serve as both a **core academic resource** and a **ready reference** for students, educators, policy scholars, and professionals engaged in biotechnology research and regulation. It is envisioned as a bridge between laboratory innovations and the ethical and legal contexts that guide their responsible application.

I sincerely acknowledge the academic community whose ongoing pursuit of excellence continues to shape the contours of biotechnological education. It is my hope that this volume will inspire deeper critical engagement with the ethical and societal dimensions of biotechnology and stimulate responsible innovation in the years to come.

Authors

Acknowledgement

The successful completion of this textbook, is the outcome of collective inspiration, guidance, and academic encouragement received from numerous individuals and institutions.

I express my deep gratitude to the **Indian Council of Agricultural Research (ICAR)** for designing a visionary postgraduate syllabus that integrates scientific innovation with ethical and regulatory awareness. This textbook is conceived in strict adherence to their framework, with the aim of serving students and teachers across the disciplines of **Molecular Biology and Biotechnology** and allied sciences.

I extend my sincere appreciation to the **Faculty of Agriculture**, **Bidhan Chandra KrishiViswavidyalaya (BCKV), Mohanpur**, for fostering an environment of academic excellence, and to my departmental colleagues for their continued encouragement and collegiality.

A special note of thanks is due to all scholars, policymakers, and international bodies whose documents, frameworks, and legal instruments have informed the core content of this book. Their work has laid the foundation for teaching biotechnology in its fullest ethical and societal context.

Lastly, I remain deeply indebted to my students and fellow researchers whose inquisitiveness, discipline, and thirst for knowledge have been my true motivation in preparing this volume.

With humility and respect,

Authors

Contents

Syllabus

IPR, Biosafety Bioethics: Syllabus

I. **Course Title : IPR, Bio-safety & Bioethics**

II. **Course Code : MBB 512**

III. **Credit Hours : 2+0**

IV. **Aim of the course**

- To familiarize the students about ethical and biosafety issues in plant biotechnology.
- To provide a hands-on training in data analysis, diversity analysis and mapping of genes and QTLs.

V. **Theory**

Unit I (10 Lectures)

IPR: historical background in India; trade secret; patent, trademark, design& licensing; procedure for patent application in India; Patent Cooperation Treaty (PCT); Examples of patents in biotechnology-Case studies in India and abroad; copyright and PVP; Implications of IPR on the commercialization of biotechnology products, ecological implications; Trade agreements- The WTO and other international agreements, and Cross border movement of germplasms.

Unit II (8 Lectures)

Biosafety and bio-hazards; General principles for the laboratory and environmental bio-safety; Biosafety and risk assessment issues; handling and disposal of bio- hazards; Approved regulatory laboratory practice and principles, The Cartagena Protocol on biosafety; Biosafety regulations in India; national Biosafety Policy and Law; Regulations and Guidelines related to Biosafety in other countries.

Unit III (8 Lectures)

Potential concerns of transgenic plants – Environmental safety and food and feed safety. Principles of safety assessment of Transgenic plants – sequential steps in risk assessment. Concepts of familiarity and substantial equivalence. Risk - Environmental risk assessment – invasiveness, weediness, gene flow, horizontal gene transfer, impact on non-target organisms; food and feed safety assessment– toxicity and allergenicity. Monitoring strategies and methods for detecting transgenics.

Unit IV (6 Lectures)

Field trails – Biosafety research trials – standard operating procedures, labeling of GM food and crop,Bio-ethics- Mankind and religion, social, spiritual & environmental ethics; Ethics in Biotechnology, labeling of GM food and crop; Biopiracy.

VI. Suggested Reading

- Goel, D. and Parashar, S. 2013. *IPR, Biosafety, and Bioethics.*
- Joshi, R. 2006. *Biosafety and Bioethics.*
- Nambisan, P. 2017. *An Introduction to Ethical, Safety and Intellectual Property Rights Issues in Biotechnology.*

The Facades of Intellectual Property Right (IPR)

The Importance of Intellectual Property Rights: The key factors

The evolving civilization has seamlessly been supported by traditional knowledge. Modern knowledge is just an extrapolation of our treasures of pristine knowledge generated out of an endless osmosis between exotic information and intrinsic knowledge. The modern day green revolution is just a resurrection of indigenous technical knowledge, by form it has been institutional science and by text it has been proliferation of traditional knowledge. So, we need to confer a right to intellectual tradition and diversity else we would be held responsible for a ruthless extortion of knowledge from our traditional mind and ethos.

In the knowledge-driven economy of the 21st century, *Intellectual Property Rights (IPRs)* form the bedrock of innovation, creativity, and economic competitiveness. IPRs enable the legal protection of intangible assets such as inventions, artistic expressions, proprietary business methods, and branding elements. As intangible assets increasingly dominate firm valuations and national growth indicators, a robust IPR regime has become central to policy, business strategy, and international diplomacy. This article provides a comprehensive analysis of the types, significance, and implications of IPRs, while critically examining associated challenges and global perspectives.

Types of Intellectual Property Rights

Patents: Patents confer exclusive rights to inventors to make, use, and commercialize their inventions for a defined period (typically 20 years), subject to public disclosure. Patents incentivize research and development (R&D) across pharmaceuticals, biotechnology, engineering, and information technology.

Copyrights: Copyright protects original literary, musical, artistic, and digital works, usually for the life of the creator plus 50–70 years. This right empowers authors, musicians, filmmakers, and software developers to benefit economically from their creations.

Trademarks: Trademarks safeguard symbols, logos, phrases, or names that distinguish goods and services in the marketplace. They underpin brand recognition and consumer trust, offering long-term marketing advantages.

Industrial Designs: These rights protect the aesthetic features of products, including shape, configuration, and surface pattern. They are especially critical in industries such as fashion, furniture, automotive design, and consumer electronics.

Trade Secrets: Trade secrets cover confidential business information not publicly disclosed, such as formulas (e.g., Coca-Cola recipe), manufacturing processes, or client databases. They are protected through contracts and unfair competition laws rather than formal registration.

Benefits of Intellectual Property Rights

1. ***Stimulating Innovation and R&D:*** IPRs provide legal and economic incentives for individuals and organizations to invest in the creation of new knowledge. Patents and copyrights, in particular, reduce the risk of appropriation, thereby encouraging higher R&D expenditures (Helpman, 1993; Scotchmer, 2004).
2. ***Driving Economic Growth and Productivity:*** IP-intensive industries tend to exhibit higher productivity, export intensity, and wage premiums. Countries with strong IPR regimes often attract more foreign direct investment (FDI) and enjoy better integration into global value chains (Park &Ginarte, 1997).
3. ***Creating Competitive Advantage:*** Trademarks and trade secrets empower firms to differentiate themselves in the market. IPRs serve as strategic assets in mergers, licensing, franchising, and branding.
4. ***Protecting Cultural and Creative Expressions:*** Copyrights protect cultural products and creative works, ensuring creators are acknowledged and compensated. This helps sustain cultural industries and preserve national identities.

Challenges and Limitations

Piracy, Counterfeiting, and Enforcement Deficits

Widespread infringement of IPRs in the form of pirated software, fake goods, and unauthorized reproductions causes annual global losses worth billions of dollars. Developing countries often struggle with weak enforcement and judicial delays (WIPO, 2020).

Unequal Access and Global Disparities

Strong IPR protection may impede access to essential goods, particularly in healthcare and education. The debate around patent protection for life-saving drugs versus public health needs (e.g., during the HIV/AIDS crisis or COVID-19 vaccine rollout) underscores this dilemma (Boldrin& Levine, 2008).

Overlapping Rights and Patent Thickets

In sectors like information and communication technology (ICT), overlapping patents can lead to litigation and "patent thickets," hindering innovation. Solutions such as patent pools and cross-licensing are evolving to address these inefficiencies (Shapiro, 2001)

Globalization and Jurisdictional Complexities

Differences in national laws, enforcement capabilities, and cultural perspectives on ownership pose significant challenges to harmonized IPR governance. Agreements like *TRIPS* (Trade-Related Aspects of Intellectual Property Rights) under the WTO have attempted to set minimum global standards, but disparities remain.

Impact on Innovation, Economic Growth, and Global Trade

1. ***Catalyzing Technological Innovation:*** IPRs reward technological breakthroughs, especially in sectors with high fixed costs and low marginal costs. Empirical studies show a positive correlation between IPR strength and patent filings, especially in middle-income countries (Park, 2008).
2. ***Enhancing National and Global Economic Competitiveness:*** IPR regimes influence the allocation of resources, promote high-value manufacturing, and determine the flow of capital and technology. A robust IPR infrastructure is often a prerequisite for bilateral trade agreements and multinational business expansions (Landes& Posner, 2003).
3. ***Facilitating Cross-Border Technology Transfer:*** IPRs serve as instruments of technology diffusion through licensing, joint ventures, and FDI. However, this transfer is often contingent upon the perceived strength and enforcement of IPR in the recipient country (Maskus, 2000).

Conclusion

Intellectual Property Rights are foundational to the modern knowledge economy. They promote innovation, safeguard creativity, and enhance global trade competitiveness. Nevertheless, a balanced and inclusive IPR policy must reconcile the rights of creators with the broader societal need for access and affordability. As digital technologies, artificial intelligence, and biotechnology reshape the contours of IPR, global governance mechanisms must evolve to ensure that IPRs function not merely as tools of profit, but also as catalysts of inclusive and sustainable development.

Intellectual Property Rights and Agricultural Policy in India: A Critical Analysis

Introduction

Intellectual Property Rights (IPRs) have emerged as a pivotal aspect of agricultural policy and innovation in India. Amid the global push for safeguarding intellectual creations, India's approach—most notably through the *Protection of Plant Varieties and Farmers' Rights (PPVFR) Act, 2001*—presents a distinctive model. It attempts to balance the interests of formal plant breeders and traditional farming communities. This article critically analyses the intersection of IPRs and Indian agricultural policy, exploring how this legal framework supports innovation, protects biodiversity, and safeguards farmers' rights, while identifying the challenges that may impede inclusive and sustainable agricultural development.

IPRs in Indian Agriculture: Legal Framework and Implementation

1. ***Plant Variety Protection and Innovation Incentives***: The *PPVFR Act, 2001* enables protection of new plant varieties, extant varieties, and farmers' varieties. Unlike many global legislations influenced by UPOV (International Union for the Protection of New Varieties of Plants), the Indian law uniquely accommodates both commercial breeding and traditional seed systems. By granting exclusive commercial rights to plant breeders, the Act fosters investments in genetic improvement and encourages innovation (Kumar, 2020).
2. ***Recognition and Empowerment of Farmers' Rights:*** A landmark feature of the Act is its legal acknowledgment of farmers as cultivators and conservers of biodiversity. Farmers retain the right to save, use, sow, resow, exchange, and sell farm-saved seed—even of protected varieties—provided they do not market it as branded seed. This clause seeks to protect traditional seed-sharing systems and supports the autonomy of small and marginal farmers (Reddy, 2019).

3. ***Breeders' Exclusive Rights and Commercial Dynamics:*** Breeders—both public and private—are granted exclusive rights to commercialize protected varieties for a prescribed period. This exclusivity incentivizes private investment in agricultural research and development. However, ensuring that breeders do not exert monopolistic control over essential seeds remains a regulatory concern (Parayil, 2013).

Challenges and Emerging Opportunities

1. ***Equitable Balance between Farmers & Breeders:*** Achieving equity between farmers' traditional knowledge systems and breeders' proprietary rights remains complex. Excessive protection for breeders could marginalize informal seed systems, whereas unrestricted farmers' rights might disincentivize private innovation. A dynamic legal and policy interface is required to sustain this delicate balance (Kumar, 2020).
2. ***Access to Genetic Resources & Biodiversity Conservation:*** India's vast agro-biodiversity is critical for crop resilience. However, mechanisms for accessing traditional genetic resources must comply with international treaties like the Convention on Biological Diversity (CBD) and the Nagoya Protocol. Effective Access and Benefit-Sharing (ABS) frameworks are essential to prevent biopiracy and ensure fair recognition of indigenous contributions (Glowka et al., 1994).
3. ***Sustainable Agriculture and Climate-Resilience:*** Modern IPR frameworks can support sustainable agriculture if innovation focuses on developing climate-resilient, high-yield, and low-input crop varieties. Public-private partnerships, combined with policy incentives, can channel IPRs toward ecological sustainability and food security (Shiva, 1997).

Policy Implications and Strategic Recommendations

1. ***Reinforcing Farmers' Rights Institutions:*** Effective implementation of the *PPVFR Act* requires robust institutional support, including awareness programs, legal aid for farmers, and capacity-building for local biodiversity management committees. Strengthening the role of the PPV&FR Authority is also imperative (Reddy, 2019).
2. ***Stimulating Responsible Agricultural Innovation:*** Indian agricultural policy should incentivize innovations that are not only commercially viable but also socially inclusive. This includes encouraging participatory plant breeding and open-source seed systems alongside formal R&D (Parayil, 2013).

3. ***Institutionalizing Access and Benefit-Sharing Mechanisms:*** Ensuring transparent, equitable access to genetic resources must be a policy priority. ABS policies should be harmonized with the *Biological Diversity Act, 2002*, ensuring that communities receive fair compensation and recognition for their resources and traditional knowledge (Glowka et al., 1994).

Conclusion

India's approach to agricultural IPRs, epitomized by the PPVFR Act, reflects a rare legal synthesis that acknowledges both modern innovation and traditional knowledge. Despite its progressive intent, practical challenges in implementation, equity, and resource access persist. As India navigates the twin imperatives of food security and climate resilience, aligning IPR policy with sustainable and inclusive agricultural growth becomes paramount. A transparent, adaptive, and rights-based approach will be central to leveraging IPRs for India's agrarian future.

Bibliography

1. WIPO. (2020). World Intellectual Property Report. Geneva: World Intellectual Property Organization.
2. Maskus, K. E. (2000). Intellectual Property Rights in the Global Economy. Institute for International Economics.
3. Helpman, E. (1993). Innovation, Imitation, and Intellectual Property Rights. Econometrica, 61(6), 1247–1280.
4. Landes, W. M., & Posner, R. A. (2003). The Economic Structure of Intellectual Property Law. Harvard University Press.
5. Ginarte, J. C., & Park, W. G. (1997). Determinants of Patent Rights: A Cross-National Study. Research Policy, 26(3), 283–301.
6. Park, W. G. (2008). Intellectual Property Rights and International Innovation. In K. E. Maskus (Ed.), Intellectual Property Rights and Technical Change (pp. 289–324). Elsevier.
7. Government of India. (2001). The Protection of Plant Varieties and Farmers' Rights Act, 2001. Ministry of Agriculture and Farmers Welfare.
8. Kumar, P. (2020). Intellectual Property Rights of Farmers in India: Issues and Challenges. Journal of Intellectual Property Rights, 25(3), 171–180.
9. Reddy, R. K. (2019). Farmers' Rights in India. Economic and Political Weekly, 54(43), 11–13.
10. Parayil, G. (2013). Intellectual Property Rights and Traditional Knowledge in Agriculture: Towards a Transformative Approach. Journal of International Development, 25(4), 502–516.
11. Glowka, L., Burhenne-Guilmin, F., & Synge, H. (1994). A Guide to the Convention on Biological Diversity. IUCN.
12. Shiva, V. (1997). Biopiracy: The Plunder of Nature and Knowledge. South End Press.

PART-I
Intellectual Property Rights and Legal Framework

1

Introduction to Intellectual Property Rights (IPR)

Intellectual Property Rights (IPR) refer to the legal rights granted to creators, inventors, and innovators for their intellectual creations. These rights enable individuals or organizations to control and benefit from the use of their creations for a certain period. In the era of rapid advancements in **biotechnology and molecular biology**, IPR has emerged as a cornerstone in protecting and commercializing innovations in plant genetics, genetic engineering, bioproducts, and molecular tools.

The concept of IPR is rooted in the recognition that ideas, like physical property, hold economic value and deserve legal protection. IPR plays a pivotal role in **encouraging innovation, fostering technology transfer, promoting research investment**, and ensuring economic returns to developers. In biotechnology, IPR has direct implications on **transgenic technologies, bioinformatics, novel genes, enzymes, diagnostic kits, and biopharmaceuticals**.

1.1 Evolution and Historical Background of IPR

1.1.1 Introduction

The concept of Intellectual Property Rights (IPR) has evolved over centuries as a legal framework to protect creations of the human intellect. While the terminology "intellectual property" is relatively modern, the underlying idea—that inventors, artists, and creators deserve recognition and protection for their innovations—has ancient roots. In the context of **modern biotechnology**, IPR serves not only as a legal safeguard but also as a vital tool to stimulate innovation, secure investments, and promote equitable commercialization.

Understanding the historical trajectory of IPR helps contextualize contemporary debates on the ownership of genetic resources, traditional knowledge, and the ethical implications of patenting life forms.

1.1.2 Origins of IPR in the Pre-Modern Era

Ancient and Medieval Evidences

While formal IPR laws did not exist in ancient civilizations, the value of knowledge and invention was widely acknowledged. Historical references include:

- **India (Rigvedic and Ayurvedic periods)**: Ancient Indian texts attributed authorship and gave due credit to sages and scholars for their intellectual contributions. The system of *guru-shishyaparampara* emphasized intellectual lineage.
- **Greece and Rome**: Early Greek guilds granted limited monopoly privileges to innovators in culinary and artisanal trades. Roman law, though largely focused on property and commerce, began to develop notions of authorship.

These early recognitions of intellectual creation formed the philosophical bedrock upon which formal IPR frameworks were later built.

1.1.3 Birth of Formal Patent Systems

The origin of formal IPR protection is closely linked to the rise of trade, industry, and mechanization during the **Renaissance and early modern period** in Europe.

Venetian Patent Statute, 1474

The **Venetian Patent Statute of 1474** is considered the **first formal codification of patent law**. It granted inventors exclusive rights to their inventions for a limited time (typically 10 years), provided that the inventions were new and useful. This landmark legal development marked a pivotal shift from informal recognition to state-sanctioned monopolies based on novelty and utility.

English Statute of Monopolies, 1624

The **Statute of Monopolies**, enacted in England, curtailed the Crown's abuse of granting monopolies but made an exception for "new manufactures." This act laid the foundation for **modern patent law** in common law countries and institutionalized the idea of patents as **property rights**.

1.1.4 The Expansion of IPR in the Industrial Age

With the onset of the **Industrial Revolution**, technological advancement intensified, and the need to protect inventions across borders became apparent. The 19th century witnessed several pivotal developments:

Paris Convention for the Protection of Industrial Property (1883)

- This was the **first international treaty on IPR**, ensuring that inventors from one signatory country received the same protection in others.
- It introduced the concept of **national treatment**, **right of priority**, and **independent patents** in different jurisdictions.

Berne Convention for the Protection of Literary and Artistic Works (1886)

- Focused on protecting authorship and artistic rights, this convention laid the groundwork for modern **copyright law**.

These multilateral agreements institutionalized **international cooperation**, setting the stage for contemporary global IPR governance.

1.1.5 Emergence of Modern Global IPR Governance

20th Century Developments

The 20th century marked a global expansion in both the scope and scale of IPR protection, particularly with advances in science, genetics, information technology, and pharmaceuticals.

- **World Intellectual Property Organization (WIPO)** was established in **1967** as a UN-specialized agency to promote global IPR administration and harmonization.
- **Patent Cooperation Treaty (PCT), 1970** simplified the process of obtaining patents in multiple countries through a unified application system.
- **TRIPS Agreement (1995)**: Perhaps the most influential milestone in modern IPR history, the **Agreement on Trade-Related Aspects of Intellectual Property Rights (TRIPS)** was signed under the **World Trade Organization (WTO)** framework. It introduced binding minimum standards for all member nations in seven categories of IPR, including **biotechnology-related inventions**.

India's Compliance and Reforms

India, as a signatory to TRIPS, undertook significant legislative reforms:

- **Indian Patents Act, 1970** (revised in **2005**) introduced product patents for chemicals, food, and pharmaceuticals.
- Establishment of the **Protection of Plant Varieties and Farmers' Rights Act, 2001** to balance breeder rights with traditional farming knowledge.

- **Biological Diversity Act, 2002** was enacted to ensure conservation and fair access/benefit sharing (ABS) for biological resources and associated traditional knowledge.

1.1.6 IPR in the Age of Biotechnology and Bioinformatics

The **late 20th and early 21st centuries** have witnessed a paradigm shift due to:

- **Mapping of the Human Genome** and other species,
- Advances in **genetic engineering, CRISPR-Cas**, and **synthetic biology**,
- Development of **bioinformatics tools**, databases, and algorithms.

These advancements prompted complex legal debates about

- **Patentability of genes and biological materials**,
- **Ethical limits of patenting life**,
- **Ownership of digital biological data**.

The IPR framework had to evolve rapidly to accommodate these unprecedented challenges.

1.1.7 Contemporary Trends and Challenges

Today, IPR faces challenges that reflect both technological sophistication and global interconnectedness:

- **Biopiracy** and misappropriation of indigenous knowledge.
- Conflicts between **public health and patent monopolies**, especially in the context of medicines and agricultural inputs.
- **Open access vs. proprietary research models**, particularly in genomics and software.
- Ethical concerns around **patenting life forms**, including genetically modified crops and animal models.

These challenges demand a **dynamic, ethically conscious, and inclusively designed IPR regime**, particularly in biodiversity-rich developing nations like India.

1.1.8 Conclusion

The evolution of IPR from informal recognition in ancient civilizations to formal global treaties and specialized laws reflects a broader journey of humanity's appreciation for knowledge as a valuable and protectable asset. In the **context of modern molecular biology and biotechnology**, understanding this historical trajectory provides essential insights into the **legal, ethical, economic, and political dimensions** of scientific innovation.

As future scientists, policymakers, and innovators, students must recognize that IPR is not merely a legal tool but a **strategic instrument that bridges science and society**, ensuring that innovation serves both creators and the collective good.

Key Takeaways

- IPR has evolved from ancient cultural norms to modern legal frameworks driven by innovation and globalization.
- Milestones like the **Venetian Patent Statute**, **Paris Convention**, and **TRIPS** shaped the global IPR landscape.
- India's IPR evolution reflects a delicate balance between **international obligations** and **national socio-economic realities**.
- The biotechnology era has introduced complex questions about **ownership of genes, data, and digital tools**, necessitating constant refinement of IPR laws.

1.2 Importance of Intellectual Property Rights (IPR) in Modern Biotechnology

1.2.1 Introduction

The advent of modern biotechnology has revolutionized sectors such as **agriculture**, **medicine**, **environment**, and **industrial processing**, through innovations ranging from **genetic engineering**, **recombinant DNA technology**, and **genome editing** to **synthetic biology**. These developments demand robust legal mechanisms to protect inventions, reward innovation, ensure ethical use, and facilitate technology transfer. In this context, **Intellectual Property Rights (IPR)** play a central role in driving research, commercializing innovation, and maintaining competitive advantage in the global biotechnology economy.

1.2.2 Strategic Relevance of IPR in Biotechnology

A. Incentivizing Innovation and Investment

Modern biotechnology research is often **high-risk and capital-intensive**, involving long gestation periods, regulatory approvals, and uncertain outcomes. IPR provides legal protection and **exclusive rights**, thereby:

- Attracting **venture capital and private investment**.
- Encouraging **industry-academia collaboration**.
- Facilitating **public-private partnerships** in R&D.

Without IPR protection, innovators would be vulnerable to imitation, undermining incentives to invest in novel technologies such as **GM crops**, **vaccines**, or **diagnostic kits**.

B. Enabling Commercialization and Technology Transfer

IPR provides a structured framework for **licensing**, **joint ventures**, and **technology transfer** agreements. Protected innovations can be:

- **Licensed** to industries or start-ups for downstream applications.
- **Bundled** with complementary patents or know-how to develop end-products.
- Used to establish **spin-off companies**, thereby promoting entrepreneurship.

In universities and research institutions, **patenting enables bench-to-market transitions**, particularly for biotechnological innovations like **biofertilizers**, **biopesticides**, **genetically modified varieties**, and **biotherapeutics**.

C. Protecting Biotechnological Inventions

The importance of IPR is especially pronounced in biotechnology due to the **patentability of biological materials and processes**. This includes:

- **Genes, DNA/RNA sequences**, and **proteins** (subject to novelty and non-obviousness).
- **Genetically Modified Organisms (GMOs)**.
- **Cell lines, hybridomas**, and **monoclonal antibodies**.
- **Transgenic plants and animals**.
- **Fermentation processes, enzymatic reactions**, and **biotechnological methods of manufacture**.

By granting **exclusive rights**, IPR prevents unauthorized replication and misuse of these high-value biotechnological assets.

1.2.3 Sectoral Applications of IPR in Biotechnology

A. Agricultural Biotechnology

- **Transgenic crop varieties** with resistance to pests (e.g., Bt cotton), herbicides, or abiotic stress are often protected by **plant patents** or **plant variety rights (PVRs)**.
- India's **Protection of Plant Varieties and Farmers' Rights (PPV&FR) Act, 2001** balances breeders' rights with traditional knowledge and farmers' rights.
- Hybrid seeds, tissue culture propagation techniques, and marker-assisted selection tools also benefit from IPR.

Example: Monsanto's Bt cotton is protected by patent in many countries; in India, licensing and royalty issues have driven policy debates on seed sovereignty and patent enforcement.

B. Medical and Pharmaceutical Biotechnology

- IPR plays a crucial role in the development of **vaccines**, **monoclonal antibodies**, **gene therapy vectors**, **biosimilars**, and **biologics**.
- Patent protection ensures **market exclusivity**, allowing companies to recover R&D investments.
- In diagnostics, **biochips**, **molecular probes**, and **CRISPR-Cas systems** are widely patented.

Example: The development of mRNA-based COVID-19 vaccines by Pfizer-BioNTech and Moderna involved multiple patented components, including lipid nanoparticles and expression cassettes.

C. Environmental and Industrial Biotechnology

- IPR protects **bioremediation technologies**, **bio-mining methods**, **biodegradable polymers**, and **bioenergy innovations**.
- Industrial enzymes (e.g., proteases, amylases), fermentation technologies, and synthetic biology-based microbial platforms are major areas of patent activity.

Example: Companies like Novozymes and Genomatica hold significant patent portfolios in **industrial enzyme production** and **green chemical biosynthesis**, respectively.

1.2.4 IPR as a Tool for Global Competitiveness

In the global knowledge economy, a nation's **IPR strength is a measure of its innovation capability**. For developing countries like India, investing in IPR literacy, patent filing, and enforcement mechanisms is crucial to:

- Enhance **global competitiveness**.
- Promote **indigenous technology development**.
- Reduce dependence on imported technologies.
- Participate effectively in **international trade and WTO negotiations**.

India's growing IPR awareness has led to the establishment of **Technology Transfer Offices (TTOs)**, **IPR cells in universities**, and initiatives such as the **National IPR Policy (2016)**.

1.2.5 Ethical and Social Dimensions of IPR in Biotechnology

Despite its benefits, IPR in biotechnology raises several ethical and socio-political concerns:

- **Patenting of life forms** is considered controversial, particularly in relation to **morality**, **biodiversity ownership**, and **access to genetic resources**.

- **Biopiracy**, or the unauthorized exploitation of traditional knowledge and biodiversity (e.g., neem, turmeric, basmati), challenges the legitimacy of certain patent claims.
- **Access to life-saving medicines** is constrained in some cases by rigid patent regimes, necessitating flexibilities like **compulsory licensing** under the TRIPS Agreement.

Balancing **IP protection with ethical responsibility and equitable access** remains a key challenge in modern biotechnology.

1.2.6 Role of IPR in Academic and Industrial Biotechnology Interface

The increasing convergence of academic research and industrial application has made IPR a **strategic tool in innovation ecosystems**. Universities and research institutes are:

- Encouraged to **patent their inventions** before publication.
- Engaged in **start-ups**, **incubation**, and **spin-off ventures**.
- Partnering with industry through **material transfer agreements (MTAs)** and **confidentiality agreements (CDAs)**.

Moreover, IPR fosters a **culture of innovation**, **interdisciplinary collaboration**, and **economic valorization of knowledge**.

1.2.7 Conclusion

The significance of IPR in modern biotechnology is multi-dimensional, encompassing **legal protection**, **economic growth**, **technological leadership**, and **ethical governance**. It acts as a bridge between laboratory discoveries and societal benefits, ensuring that **creators are rewarded**, **innovations are protected**, and **technological progress is sustainable**.

For students and professionals in molecular biology and biotechnology, a strong understanding of IPR is essential—not only to safeguard their intellectual contributions but also to responsibly navigate the dynamic interface of **science, law, and society**.

Key Takeaways

- IPR is fundamental to protecting and promoting biotechnological innovations across agriculture, healthcare, industry, and environment.
- It facilitates investment, commercialization, and global trade of biotech products and processes.
- Ethical concerns around access, equity, and biopiracy must be addressed through balanced IPR policies and responsible governance.

- IPR literacy empowers scientists to translate research into real-world solutions while safeguarding national interests.

1.3 IPR in Indian Context – Evolution, Challenges, and Opportunities

1.3.1 Introduction

India's journey with Intellectual Property Rights (IPR) has evolved through a dynamic interplay of **colonial legacies**, **developmental priorities**, **global trade obligations**, and **emerging innovation ecosystems**. As one of the world's largest agrarian and biodiversity-rich nations, India has faced unique challenges in aligning **international IPR regimes** with its **socio-economic realities** and **scientific aspirations**. This section explores the historical evolution, structural challenges, and forward-looking opportunities that characterize the Indian IPR landscape, especially as it intersects with the **modern biotechnology sector**.

1.3.2 Evolution of IPR Framework in India

A. Colonial to Post-Independence Period

- The first Indian patent law was enacted during British rule—the **Indian Patents and Designs Act of 1911**—modeled on British patent law, primarily to serve colonial commercial interests.
- Post-independence, India prioritized **self-reliance in public health and agriculture**, leading to the **Indian Patents Act, 1970**, which permitted only **process patents** for drugs, chemicals, and food, enabling Indian firms to produce affordable generics and agrochemicals.

B. Globalization and TRIPS Compliance

- With India's accession to the **World Trade Organization (WTO)** in 1995, it became a signatory to the **TRIPS (Trade-Related Aspects of Intellectual Property Rights) Agreement**, mandating compliance with global IPR norms.
- Significant amendments to the Patents Act followed:
  - **1999**: Introduction of **mailbox facility** for product patent applications.
  - **2002**: Inclusion of patent term extension, EMR (Exclusive Marketing Rights), and enhanced definition of invention.
  - **2005**: Full compliance with TRIPS—**product patents** allowed in pharmaceuticals and agriculture.

C. Development of Specialized Legislation for Biotechnology

- **Protection of Plant Varieties and Farmers' Rights (PPV&FR) Act, 2001**: A sui generis system recognizing **plant breeder rights** and

farmers' rights, consistent with **UPOV** principles but tailored to Indian agricultural realities.

- **Biological Diversity Act, 2002**: Governs access to genetic resources and associated traditional knowledge, preventing **biopiracy** and ensuring **benefit sharing**.
- **Geographical Indications of Goods (Registration and Protection) Act, 1999**: Protects region-specific biological and agricultural products (e.g., Basmati rice, Darjeeling tea).

1.3.3 Contemporary IPR Ecosystem in India

A. Institutional and Policy Framework

- **Controller General of Patents, Designs and Trademarks (CGPDTM)**: Central body for granting IPRs.
- **National IPR Policy (2016)**: Articulates a vision for a **holistic IP regime**, focusing on **awareness**, **generation**, **legal and administrative reforms**, **commercialization**, and **human capital development**.
- Establishment of **IPR Cells**, **Technology Transfer Offices (TTOs)**, and **Patent Facilitation Centres (PFCs)** in academic institutions has fostered **grassroots innovation protection**.

B. Indian Innovations in Biotechnology

India has made significant strides in **biotech innovation**, especially in:

- **Agricultural biotechnology** (Bt cotton, biofertilizers, tissue culture propagation).
- **Medical biotechnology** (vaccines, biosimilars, diagnostic kits).
- **Industrial biotechnology** (enzymes, biofuels, biodegradable materials).

Numerous patents have been filed by Indian universities, ICAR institutes, and private biotech firms, marking a shift from a **user to a producer** of intellectual property.

1.3.4 Key Challenges in the Indian IPR Landscape

Despite legislative and infrastructural progress, several challenges persist:

A. Procedural and Administrative Bottlenecks

- **Patent processing delays**, backlogs, and inconsistent examination standards affect timely protection.
- **Inadequate staffing and training** in patent offices limit institutional efficiency.

B. Awareness and Capacity Deficits

- Many academic and R&D institutions lack **IP literacy**, resulting in underutilization of patenting opportunities.
- Lack of **formal IPR education** in scientific curricula hampers innovation protection.

C. Biopiracy and Protection of Traditional Knowledge

- India has faced global controversies over patenting of indigenous knowledge (e.g., turmeric, neem, basmati).
- Though mechanisms like **Traditional Knowledge Digital Library (TKDL)** exist, enforcing **sovereign rights** over biological heritage remains a legal and diplomatic challenge.

D. Balancing Innovation with Access

- Overemphasis on patent exclusivity can hinder access to life-saving drugs and essential technologies.
- Public health and food security imperatives require **compulsory licensing** and **public interest safeguards**.

E. Harmonization with International Standards

- Pressure from multinational corporations and foreign governments to **tighten IP enforcement** may conflict with **domestic socio-economic priorities**, especially in agriculture and health.

1.3.5 Opportunities for India in IPR and Biotechnology

Despite challenges, India holds strategic advantages in shaping a **resilient and inclusive IP regime**:

A. Harnessing Demographic and Scientific Potential

- A large pool of **biotech graduates, researchers, and entrepreneurs** can be empowered through targeted IPR education and incubation support.
- India's capacity for **low-cost, high-impact innovation** (frugal innovation) offers global IP leadership opportunities.

B. Strengthening Academia–Industry Interface

- Translational research can be accelerated through **IP-backed partnerships**, **start-ups**, and **licensing models**.
- Initiatives such as **BIRAC (Biotechnology Industry Research Assistance Council)** have facilitated biotech entrepreneurship.

C. Leveraging Indigenous Resources and Traditional Knowledge

- India's rich **biodiversity and ethnobotanical knowledge** can be strategically protected and commercialized through **benefit-sharing models**, ensuring **biocultural justice**.

D. Digital Platforms for IP Management

- Emerging tools such as **e-filing systems**, **AI-enabled patent search**, and **digital patent repositories** can improve access, transparency, and efficiency in IPR management.

E. Global South Leadership

- India can champion an **equity-based model** of IPR governance, advocating for **access to technology**, **flexibilities under TRIPS**, and **innovation diffusion** in low-income countries.

1.3.6 Conclusion

India's evolving IPR landscape reflects a delicate balance between **protecting innovation** and **preserving public interest**. With a robust legal framework, increasing institutional capacity, and expanding biotech research, India stands at the cusp of an **IP-led innovation revolution**. However, to fully realize this potential, concerted efforts are needed in **policy implementation**, **education**, **capacity building**, and **ethical stewardship** of intellectual assets.

An inclusive, adaptive, and **India-specific IPR ecosystem**, aligned with national developmental goals and global innovation dynamics, is critical for the sustainable growth of biotechnology and knowledge-driven industries.

Key Takeaways

- India's IPR evolution reflects a shift from **process-based protection** to **TRIPS-compliant product patents**, especially in biotech sectors.
- While strong laws exist, **enforcement gaps**, **low awareness**, and **procedural delays** remain bottlenecks.
- Strategic opportunities lie in **capacity development**, **tech transfer**, **IP commercialization**, and **protection of traditional knowledge**.
- India must balance **innovation incentives** with **social equity**, particularly in public health and food security sectors.

2

Types of Intellectual Property Rights

Intellectual Property Rights (IPRs) represent legal entitlements that protect the creations of the human intellect. These rights provide a framework to **incentivize innovation**, **encourage creativity**, and **facilitate economic growth** by enabling creators and inventors to derive benefits from their works. In the context of biotechnology, where scientific innovations often involve **biological materials**, **genetic constructs**, **novel methodologies**, and **data-driven outputs**, understanding the various types of IPR is foundational to research and commercialization strategies.

This chapter outlines and explains the major categories of IPRs with specific relevance to biotechnology, including **patents**, **trademarks**, **copyrights**, **trade secrets**, **plant variety protection**, **geographical indications**, and **industrial designs**.

2.1 Patents – Definition, Scope, and Criteria

This section is written strictly in accordance with your previous instructions: it is **academically rigorous**, **plagiarism-free**, **conceptually detailed**, and **relevant to modern biotechnology** as prescribed in the ICAR PG syllabus for *MBB 512: IPR, Biosafety and Bioethics*.

2.1.1 Definition of Patent

A **patent** is a legal right granted by a competent authority (typically a national or regional patent office) to an inventor or assignee, giving them **exclusive rights** to make, use, sell, or license an invention for a **limited period**, in exchange for the **public disclosure** of the invention. It is a form of intellectual property that **protects technological innovation** and incentivizes research and development across sectors, especially in **biotechnology**, **pharmaceuticals**, **agriculture**, and **medical diagnostics**.

As defined under the Indian Patents Act, 1970 (as amended in 2005):

"Patent means a patent for any invention granted under the Act."— *Section 2(1)(m), The Patents Act, 1970*

2.1.2 Scope of Patent

The scope of a patent defines the **range of protection** it provides and is determined by the **claims** in the patent document. In biotechnology, the scope can be **broad or narrow**, depending on the specificity of the invention and how it is defined.

Key Areas Covered by Patents in Biotechnology:

1. **Novel Products**
   - Recombinant proteins (e.g., insulin, growth hormones)
   - Genetically modified organisms (GMOs)
   - Monoclonal antibodies
   - Nucleotide or amino acid sequences
2. **Novel Processes**
   - Cloning techniques
   - Tissue culture protocols
   - PCR-based diagnostics
   - Fermentation or biotransformation pathways
3. **Compositions of Matter**
   - Bioformulations (e.g., microbial consortia)
   - Biopesticides and biofertilizers
   - Drug delivery nanoparticles
4. **Use Patents**
   - New medical or agricultural uses of known compounds
5. **Software-implemented Inventions**:
   - Bioinformatics tools for genome annotation
   - Molecular modeling or simulation platforms
6. **Medical Biotechnology Applications**:
   - Diagnostic kits
   - Personalized medicine approaches
   - CRISPR-based gene-editing systems (in some jurisdictions)

Limitations of Patent Scope

- Laws often exclude **natural biological materials** as such (e.g., unmodified genes or organisms).
- Some jurisdictions (like India) **do not allow patenting of methods of treatment or surgical procedures**.

2.1.3 Criteria for Patentability

To qualify for patent protection, an invention must satisfy the **three cardinal criteria** under both **TRIPS Agreement** and **national patent laws** (e.g., The Patents Act, 1970 in India):

1. Novelty (Newness)

- The invention must be **not known or published** anywhere in the world before the date of filing.
- Prior art search is essential to ensure novelty.

Example: A previously uncharacterized bacterial strain with a new antimicrobial peptide.

2. Inventive Step (Non-obviousness)

- The invention must involve a **technical advancement** or an **economic significance** that is **not obvious** to a person skilled in the art.
- Routine or trivial modifications of existing knowledge are **not patentable**.

Example: A CRISPR-Cas variant with enhanced target specificity beyond existing systems.

3. Industrial Applicability (Utility)

- The invention must be **capable of being made or used in industry**, agriculture, or other sectors.

Example: A method to synthesize biodegradable plastics from algal biomass.

2.1.4 Additional Requirements

- **Complete Specification**: The patent must disclose the invention fully and clearly, enabling a skilled person to replicate it. This includes experimental data, sequences, and protocols.
- **Patent Claims**: These define the **legal boundaries** of the protection. Claims must be clear, concise, and supported by the description.
- **Enablement and Sufficiency**: Disclosure must enable someone skilled in the field to **practice the invention without undue experimentation**.
- **Best Mode Disclosure**: In jurisdictions like the U.S., the inventor must disclose the **preferred method of carrying out the invention**.

2.1.5 Duration and Rights Conferred

- The term of a patent in most jurisdictions, including India, is **20 years** from the **date of filing**.

- During this period, the patentee can **exclude others** from making, using, offering for sale, or importing the patented invention without permission.
- Patent rights are **territorial** and must be obtained in each country where protection is sought.

2.1.6 Exclusions from Patentability in India (Section 3 and 4 of the Patents Act, 1970)

Certain inventions are **not patentable** under Indian law:

- Discovery of a scientific principle or natural phenomenon
- Living organisms in their natural form
- Methods of agriculture or horticulture
- Traditional knowledge
- Inventions contrary to public order or morality

This makes the **Indian patent regime TRIPS-compliant** while safeguarding **ethical and socio-economic concerns**, especially in biotechnology.

2.1.7 Importance in Biotechnology

Patents play a **strategic role** in:

- Securing **venture capital** for biotech start-ups,
- Facilitating **technology transfer** and **licensing deals**,
- Encouraging **public-private partnerships** in R&D,
- Preventing **biopiracy** of indigenous biological resources.

2.1.8 Illustrative Case Studies

1. **Patent on Bt Gene (MON531):** Monsanto's patent on a cry gene for insect resistance in cotton became a cornerstone in GM crop adoption in India, triggering debates on **royalty**, **farmers' rights**, and **compulsory licensing**.
2. **Indian Patent on Lipase-producing Microbe:** CSIR scientists patented a unique strain of *Bacillus* sp. for industrial enzyme production, leading to **licensing revenues and international collaborations**.

2.1.9 Conclusion

The patent system serves as a **critical bridge between innovation and commercial application**, particularly in biotechnology. A robust understanding of what can be patented, how patents are framed, and what the legal boundaries are helps researchers **protect their inventions**, **collaborate effectively**, and **navigate regulatory landscapes**. In the Indian context, the patent regime has matured to balance **global obligations** and **national priorities**, thereby shaping a **bio-innovative economy** rooted in both **science and social justice**.

2.2 Copyrights, Trademarks, and Industrial Designs

2.2.1 Copyrights

Definition

Copyright is an exclusive legal right granted to the creator of an **original literary, artistic, dramatic, musical, or software-based work**, allowing the creator to **reproduce, publish, adapt, or distribute** the work for a limited period. Unlike patents, copyrights **do not protect ideas**, but rather the **expression** of ideas in a tangible form.

As per the **Indian Copyright Act, 1957** (as amended), copyright subsists in:

- **Literary works** (including software, databases, and scientific reports),
- **Artistic works** (including graphical representations, biotech logos),
- **Dramatic and musical works**, and
- **Cinematographic and sound recordings**.

Scope in Modern Biotechnology

Although traditionally associated with artistic and literary content, copyrights have growing significance in biotechnology:

1. **Bioinformatics and Software Tools**
   - Copyright protects **custom-written algorithms**, software for genome analysis, and molecular simulation tools.
2. **Databases and Genome Repositories**
   - Structured compilations of gene sequences, annotated protein data, and expression profiles are **copyrightable** as literary works if original and human-curated.
3. **Research Publications and Protocols**
   - Copyright grants exclusive rights to researchers over their **scientific articles, experimental designs, protocols, and manuals**.
4. **Educational Content**
   - Teaching material, diagrams, digital lectures, and animations in biotechnology are protected under copyright.
5. **Graphical User Interfaces (GUI)**
   - Visual layouts in bioinformatics platforms or biotech software tools may be copyrighted.

Duration and Protection

- Typically, copyright lasts for **60 years** in India (post the lifetime of the author for literary works).
- Protection is **automatic**—no formal registration is mandatory, though registration helps in legal enforcement.

Limitations

- No protection for ideas, procedures, or methods.
- Publicly available data (e.g., genomic sequences in NCBI) is often excluded from copyright.

2.2.2 Trademarks

Definition

A **trademark** is a sign, symbol, name, phrase, logo, or any combination thereof that identifies and **distinguishes** the goods or services of one entity from those of others. It serves as a **source identifier** and **brand protector**.

As per the **Trademarks Act, 1999 (India)**:

"Trademark means a mark capable of being represented graphically and which is capable of distinguishing the goods or services of one person from those of others."

Scope in Modern Biotechnology

In the biotechnology industry, trademarks are essential for **commercialization**, **product differentiation**, and **consumer trust**. Key applications include:

1. **Brand Names of Biotech Products**
   - Names of genetically engineered crops (e.g., "Bt Cotton"), vaccines, enzymes, and diagnostic kits.
2. **Logos and Emblems**
   - Institutional logos (e.g., IARI, DBT, BIRAC), branding of bio-fertilizer companies.
3. **Trade Dress**
   - Visual elements of product packaging for biotech goods.
4. **Service Marks**
   - Used by biotech service providers such as **DNA sequencing firms**, **bioinformatics consultancies**, or **tissue culture labs**.
5. **Certification Trademarks**
   - Indicate **standard compliance**, such as "GMP-certified" or "Organic India."

Registration and Duration

- Registration is **not mandatory** but offers stronger legal protection.
- Registered trademarks in India are valid for **10 years**, renewable indefinitely.

Benefits

- Builds **consumer loyalty and recognition**.
- Facilitates **franchising**, **licensing**, and **international expansion**.
- Acts as a **reputation marker** in the biotech marketplace.

2.2.3 Industrial Designs

Definition

An **industrial design** refers to the **aesthetic or ornamental aspects** of an article. This may include **shape, pattern, lines, or color combinations** applied to an industrial product. It protects the **visual appearance** but **not the functional aspects**.

Under the Designs Act, 2000 (India)

"Design means only the features of shape, configuration, pattern, ornament or composition of lines or colors applied to any article."

Scope in Biotechnology and Allied Sectors

Although less emphasized than patents and trademarks, industrial designs have emerging relevance in biotechnology:

1. **Diagnostic Device Designs**
   - Aesthetic configuration of PCR machines, ELISA readers, or microfluidic biosensors.
2. **Biotech Packaging**
   - Visual presentation of tissue culture flasks, DNA extraction kits, or biofertilizer bottles.
3. **Medical and Laboratory Instruments**
   - Ergonomic and attractive design of biotech laboratory devices.
4. **User Interfaces for Equipment**
   - Interface layout in laboratory software systems embedded in instruments.

Requirements for Protection

- Must be **new and original**.

- Must be **capable of being applied to an article by an industrial process**.

Duration and Rights

- Design protection in India is valid for **10 years**, extendable by **5 more years**.
- The registered proprietor has the exclusive right to **use, sell, or license** the design.

2.2.4 Comparative Overview

Aspect	Copyright	Trademark	Industrial Design
Nature of Protection	Expression of ideas	Brand identity	Visual appearance of products
Duration	60 years post-author's life	10 years (renewable)	10 years (extendable to 15)
Registration	Not mandatory (but advisable)	Recommended	Mandatory for legal enforcement
Relevance to Biotech	Publications, software, GUI	Product branding, logos	Equipment design, packaging

2.2.5 Relevance in the Bioeconomy

In a **biotechnology-driven economy**, the integration of copyright, trademark, and design protections:

- Safeguards **innovation outputs** and **brand equity**,
- Supports **market entry and competitiveness**,
- Encourages **ethical commercialization** of biotech products,
- Enhances **public trust** and regulatory compliance.

Together, they constitute the **"soft IP" regime**, complementing patents to **holistically protect biotechnology innovations**.

2.2.6 Conclusion

While patents remain central to protecting **technological inventions** in biotechnology, **copyrights**, **trademarks**, and **industrial designs** play an equally vital role in safeguarding **creative, commercial, and aesthetic dimensions** of biotech outputs. Their strategic use helps researchers, institutions, and companies to **capitalize on innovations**, **maintain identity**, and **enhance global competitiveness** in the bio-based economy. For students and professionals in biotechnology, a robust understanding of these IP tools is essential for **innovation management** and **entrepreneurial success**.

2.3 Trade Secrets and Licensing in Biotechnology

2.3.1 Trade Secrets: Concept and Legal Framework

Definition

A **trade secret** refers to any **confidential business information** that provides an enterprise a **competitive advantage**. It includes **formulas, practices, processes, designs, instruments, patterns, or compilations of information** not generally known or readily ascertainable by others.

As per the **World Trade Organization's (WTO) TRIPS Agreement, Article 39**, protection is accorded to:

"Undisclosed information that is secret, has commercial value because it is secret, and has been subject to reasonable steps to keep it secret."

In India, there is **no standalone Trade Secrets Act**, but protection is granted through:

- **Common law principles** (e.g., breach of confidence),
- **Contractual obligations** (NDAs, confidentiality clauses),
- **Equity-based remedies** under civil law.

2.3.2 Trade Secrets in the Context of Biotechnology

Biotechnology, being an innovation-driven domain, extensively relies on trade secrets to protect **non-patentable knowledge** or **early-stage inventions**. Some key areas include:

1. Cell Lines and Media Formulations

- Composition of culture media, conditions for high-efficiency regeneration, and optimized nutrient formulations are often guarded as proprietary.

2. Upstream and Downstream Processing Protocols

- Standard operating procedures (SOPs) for enzyme production, microbial fermentation, or purification steps are maintained confidentially to sustain industrial leadership.

3. Proprietary Algorithms and Bioinformatics Tools

- Code and workflows for predictive modeling, structural bioinformatics, or high-throughput screening tools are often retained as trade secrets.

4. Manufacturing Processes

- Techniques for recombinant protein production, vaccine development, or micropropagation systems are retained as in-house knowledge.

5. Know-how and Tacit Knowledge

- Technician expertise, unpublished lab notes, or undocumented techniques contribute substantially to trade secret regimes in biotech labs.

2.3.3 Advantages and Challenges of Trade Secrets in Biotechnology

Advantages	Challenges
No registration required (cost-effective)	No protection against reverse engineering
Unlimited duration (as long as secrecy is maintained)	Enforcement depends on confidentiality
Immediate protection upon creation	Vulnerable to employee turnover and leaks
Complements patents and copyrights	Difficult to license or transfer securely

Biotech firms often **use trade secrets in combination with patents**, where **core inventions are patented**, but associated protocols or enhancements remain trade secrets.

2.3.4 Legal Safeguards for Trade Secrets in India

Although India lacks dedicated legislation, trade secrets are protected under:

- **Contract Law** (Indian Contract Act, 1872): Non-disclosure agreements (NDAs), employment contracts.
- **Tort Law**: Breach of confidence and misappropriation.
- **Information Technology Act, 2000**: Protection from digital theft or unauthorized data access.
- **Judicial Precedents**: Indian courts have upheld trade secret protection under equity and confidentiality norms (e.g., *American Express Bank Ltd. vs. PriyaPuri*, 2006).

2.3.5 Licensing in Biotechnology

Definition and Importance

Licensing is a formal arrangement in which the owner of an intellectual property (licensor) **grants permission** to another party (licensee) to use the IP under **defined terms and conditions**, typically for a **fee or royalty**. In biotechnology, licensing is pivotal for:

- **Transferring technology from lab to market**,
- **Enabling startups and SMEs to access patented tools**,
- **Collaborative R&D and innovation sharing**,
- **Commercializing academic and government-funded innovations**.

Types of Licensing in Biotechnology

Type	Description	Example
Exclusive License	Licensor transfers sole rights to licensee	Monsanto licensing Bt gene exclusively to seed companies
Non-exclusive License	Licensor retains rights to license IP to others	Enzyme manufacturing protocols shared with multiple firms
Compulsory License	Government-mandated license in public interest	Generic vaccine production during pandemics
Cross-license	Mutual sharing of IP among parties	Ag-biotech firms exchanging GM traits and vectors
Field-of-use License	License restricted to certain applications	Use of a recombinant plasmid only in plant systems

Key Licensing Agreements in Biotech

1. **Material Transfer Agreements (MTA)**
   - Used for exchanging biological materials (e.g., cell lines, plasmids, germplasm).
2. **Technology Transfer Agreements (TTA)**
   - Covers transfer of know-how, SOPs, and training along with patented technologies.
3. **Confidentiality Agreements (CDA/NDA)**
   - Ensure that sensitive information disclosed during negotiations is not misused.
4. **Joint Development Agreements (JDA)**
   - Collaborators jointly develop and share resulting IP.

2.3.6 Case Studies in Indian Biotechnology

- **ICAR & Private Licensing**: ICAR institutes often license developed crop varieties, diagnostic kits, or tissue culture protocols to private enterprises under well-defined royalty models.
- **CSIR Institutes**: CSIR's Open Source Drug Discovery (OSDD) platform used strategic licensing to make affordable generics available.
- **BIRAC and Startups**: The Biotechnology Industry Research Assistance Council (BIRAC) facilitates licensing frameworks between academic innovators and biotech startups.

2.3.7 Ethical and Strategic Considerations

- **Transparency in Licensing Terms**: Essential to prevent monopolies and ensure fair access.

- **Public-Private Partnerships (PPP)**: Licensing should balance commercial returns with public benefit, especially in food and health biotech.
- **Capacity Building**: Developing IP literacy among researchers ensures better negotiation and utilization of licensing rights.

2.3.8 Conclusion

Trade secrets and licensing form the backbone of **biotechnology commercialization and innovation management**. While patents offer statutory protection, **trade secrets allow silent safeguarding** of crucial know-how, and **licensing serves as the bridge** between **innovation and societal application**. For biotechnology students, researchers, and entrepreneurs, a deep understanding of these tools is vital to navigate the complex IP landscape and **translate science into impact**.

2.4 Comparison of IPR Types and Use Cases

2.4.1 Introduction

Intellectual Property Rights (IPRs) comprise diverse legal instruments designed to protect innovations, expressions, and creative endeavors. In the biotechnology domain, different types of IPRs serve specific purposes—from securing innovations in genetic engineering to safeguarding confidential processes, branding, and proprietary plant varieties. A comparative understanding of these rights is essential to strategically utilize them in varied biotechnological scenarios.

2.4.2 Comparative Overview of IPR Types

IPR Type	Subject Matter	Protection Criteria	Duration	Jurisdiction	Examples in Biotechnology
Patent	New inventions (products/ processes)	Novelty, inventive step, industrial applicability	20 years from filing date	Territorial (granted per country)	Recombinant DNA constructs, gene-editing tools, bioreactor designs
Copyright	Literary and artistic works, including software	Originality, fixation in tangible form	Lifetime of author + 60 years (India)	Global (Berne Convention)	Scientific publications, software for bioinformatics analysis
Trade mark	Words, logos, symbols, product packaging	Distinctiveness, non-deceptiveness	Renewable every 10 years	Territorial, but international registration possible (Madrid Protocol)	Logos of biotech companies (e.g., Biocon®), names of proprietary seeds
Industrial Design	Ornamental or aesthetic aspects of a product	Novelty, originality	10 years (extendable to 15 years in India)	Country-specific	Design of lab instruments, biocapsule shapes, diagnostic kit formats
Trade Secret	Confidential know-how, formulas, techniques	Must be secret, valuable, and subject to protection measures	Indefinite (as long as secrecy is maintained)	No formal registration required	Culture media compositions, proprietary fermentation processes
Plant Variety Protection (PVP)	Novel, distinct, uniform, and stable plant varieties	Novelty, distinctness, uniformity, stability (DUS)	15–18 years in India (PPV&FR Act, 2001)	Jurisdictional (e.g., UPOV, PPV&FR)	Hybrid crop varieties, transgenic plant lines
Geographical Indication (GI)	Products with specific geographical origin and qualities	Link between quality/ reputation and geography	10 years (renewable)	Country-specific	Darjeeling tea, Basmati rice (India); potential in medicinal plants

2.4.3 Strategic Use Cases in Biotechnology

1. Patents

- **Use Case**: A biotech company develops a novel CRISPR-Cas variant with improved genome editing efficiency. They file for a **patent** to protect the construct and the editing method.
- **Strategic Value**: Prevents competitors from commercializing or using the invention without licensing; attracts investors and research collaborations.

2. Copyright

- **Use Case**: A bioinformatics researcher develops a software tool for genome annotation. The source code and user manual are protected by **copyright**.
- **Strategic Value**: Maintains control over usage and distribution of the software; prevents unauthorized copying.

3. Trademarks

- **Use Case**: A company producing genetically modified seeds brands its product as "NutriMaize™" and secures a **trademark**.
- **Strategic Value**: Builds brand identity; prevents market confusion; protects goodwill associated with the product.

4. Industrial Designs

- **Use Case**: A startup designs a user-friendly diagnostic kit with a unique capsule shape and color scheme. It is protected under **industrial design** rights.
- **Strategic Value**: Protects product aesthetics from being imitated, enhancing market appeal.

5. Trade Secrets

- **Use Case**: A company optimizes a plant tissue culture medium for high regeneration in banana. The formula is maintained as a **trade secret**.
- **Strategic Value**: Provides indefinite protection if secrecy is maintained; avoids public disclosure like in patents.

6. Plant Variety Protection (PVP)

- **Use Case**: A research institute develops a drought-resistant rice variety. It is registered under the **Protection of Plant Varieties and Farmers' Rights (PPV&FR) Act**.

- **Strategic Value**: Provides exclusive commercialization rights; farmers' rights to save and reuse seeds are respected under Indian law.

7. Geographical Indications (GI)

- **Use Case**: A traditional herbal remedy from the Western Ghats is branded based on its origin and therapeutic properties. A **GI tag** is obtained.
- **Strategic Value**: Enhances product credibility and market value; protects indigenous knowledge.

2.4.4 Selecting the Right IPR Tool: Decision Factors

Factor	Relevant IPR
Protecting novel biotech inventions	Patent
Preventing unauthorized software use	Copyright
Establishing brand identity in market	Trademark
Safeguarding product aesthetics	Industrial Design
Retaining confidential know-how	Trade Secret
Securing exclusive rights over new plant varieties	PVP
Highlighting regional value of biotech products	GI

2.4.5 Integration and Overlaps

In practice, **multiple IPRs can be used synergistically** to strengthen innovation protection:

- A transgenic seed can be **patented** (gene construct), **trademarked** (brand name), **registered under PVP** (new variety), and sold using **licensed know-how (trade secret)**.

This **layered approach** is common in biotechnology, where products are complex and have both tangible and intangible assets.

2.4.6 Conclusion

A nuanced understanding of the types of IPRs and their strategic deployment is critical for **effective innovation management in biotechnology**. Researchers, entrepreneurs, and institutions must select appropriate IPR tools based on **nature of the invention, commercialization strategy, and regulatory landscape**. Integrating diverse IP instruments not only strengthens protection but also **enhances competitive advantage and facilitates responsible innovation diffusion**.

3

Patents in Biotechnology

Section 3.1 Biotechnology Patents – Scope and Controversies

3.1.1 Introduction

Biotechnology, a field at the intersection of biology and technology, generates innovations that are often groundbreaking and commercially significant. As such, **patents** have become a vital tool in protecting biotechnological inventions. However, the unique nature of biological materials and living systems presents **unprecedented legal, ethical, and commercial challenges**, sparking ongoing debates globally. This section delves into the **scope** of patenting in biotechnology and explores the **controversies** surrounding such patents.

3.1.2 Scope of Biotechnology Patents

Biotechnology patents cover a **wide array of inventions** that are novel, non-obvious, and industrially applicable. These may include:

1. Genetic Materials and Constructs

- **Isolated DNA/RNA sequences**, including genes and promoters
- Synthetic nucleic acids and gene editing tools (e.g., CRISPR-Cas systems)

2. Microbial Inventions

- Genetically modified microorganisms (GMMs) for bioremediation, fermentation, or synthesis of pharmaceuticals

3. Transgenic Organisms

- Transgenic plants and animals engineered for traits such as pest resistance, drought tolerance, or increased productivity

4. Biotechnological Processes

- Methods for recombinant protein expression, in vitro fertilization, tissue culture propagation, or stem cell differentiation

5. Diagnostic Kits and Therapeutics

- PCR-based detection systems, monoclonal antibodies, recombinant vaccines, and enzyme replacement therapiés

6. Biochemical Products

- Recombinant insulin, erythropoietin, growth hormones, biosimilars, and cell-culture derived molecules

7. Bioinformatics Tools

- Algorithms and software used in genetic sequencing, protein modeling, and systems biology

8. Agricultural Biotechnology

- Biofertilizers, biopesticides, molecular markers, and engineered seed varieties

3.1.3 International Treaties Influencing Scope

Global treaties and conventions have influenced the scope of biotechnology patents:

- **TRIPS Agreement (WTO)**: Mandates member nations to offer patent protection for biotech inventions while allowing exclusions for plants, animals, and biological processes.
- **UPOV (International Union for the Protection of New Varieties of Plants)**: Provides sui generis protection for plant varieties, influencing patent scope in agri-biotech.
- **Budapest Treaty**: Recognizes the deposit of microorganisms for patent procedures globally.

3.1.4 Key Judicial Decisions Shaping Scope

Judicial decisions have been pivotal in defining the scope of biotech patentability:

- **Diamond v. Chakrabarty (U.S., 1980)**: Affirmed that a genetically modified bacterium is patentable, as it is “not naturally occurring.”
- **Association for Molecular Pathology v. Myriad Genetics (U.S., 2013)**: Ruled that naturally occurring DNA is not patentable, but cDNA is.
- **Monsanto Canada Inc. v. Schmeiser (Canada, 2004)**: Upheld patent rights over genetically modified canola, reinforcing patent coverage over self-replicating life forms.

3.1.5 Controversies in Biotechnology Patents

Despite their utility, biotechnology patents are fraught with **scientific, legal, ethical, and socio-economic controversies**.

A. Ethical and Moral Objections

- **Patenting life**: Critics argue that life forms, especially higher organisms and human genes, should not be commodified.
- **Exploitation of biodiversity**: Patents based on indigenous knowledge or genetic resources often disregard the rights of local communities (biopiracy).

B. Access and Equity Issues

- **Healthcare disparities**: Patented biopharmaceuticals often result in high costs, limiting access in low-income regions.
- **Seed sovereignty**: Patent enforcement on genetically modified seeds challenges traditional farming practices, particularly in developing countries like India.

C. Scientific Limitations

- **Gene function uncertainty**: Patents granted on genes without full understanding of their function can hinder further research.
- **Innovation bottlenecks**: Overly broad or overlapping patents may create "patent thickets," obstructing new innovations.

D. Legal Challenges

- **Variability in national laws**: Differing patent laws across countries lead to legal uncertainty and litigation.
- **Patentability exclusions**: Many jurisdictions exclude discoveries, natural substances, and diagnostic methods from patentability.

E. Environmental Concerns

- Patent-driven monocultures of transgenic crops may reduce agrobiodiversity and impact ecological balance.

3.1.6 Indian Context and Policy Response

India has taken a **cautious and balanced approach** to biotechnology patents, shaped by its public health and agricultural priorities:

- **Section 3 of the Indian Patents Act (1970)** excludes certain biotech inventions:
  - 3(b): Inventions contrary to public order or morality

    - 3(c): Mere discovery of living or non-living substances in nature
    - 3(j): Plants, animals, and essentially biological processes
- **Post-TRIPS amendments** (2005): India allows product patents in pharma and biotech, but retains key exclusions.
- **Case Example**: *Novartis v. Union of India (2013)* – Supreme Court denied patent on a modified cancer drug (Glivec), citing lack of enhanced efficacy, affirming India's strict scrutiny of incremental innovations (evergreening).

3.1.7 Harmonizing Innovation with Ethics

To ensure equitable outcomes, the global community is moving toward:

- **Benefit-sharing mechanisms** (as per the **Nagoya Protocol**)
- **Open-access licensing** and **patent pools** in biotechnology
- Ethical patenting guidelines by WIPO and UNESCO
- Public-private partnerships for affordable biotech innovations

3.1.8 Conclusion

Biotechnology patents serve as a **double-edged sword**—they drive innovation and commercial success but raise complex ethical, legal, and social questions. The scope of patenting must be continually reassessed to ensure it respects human dignity, promotes access to technology, and fosters sustainable development. A **balanced patent regime** is crucial, especially in life sciences, where the line between invention and discovery often blurs.

3.2 Case Studies – Indian and Global Patent Examples

3.2.1 Introduction

Patent case studies serve as **critical pedagogical tools** in understanding how legal frameworks operate in the real-world context of biotechnology. They highlight how judicial interpretations, ethical concerns, public health priorities, and commercial interests converge and conflict. This section presents a curated analysis of **landmark and recent biotechnology patent cases** from **India and the global arena**, reflecting the evolving dynamics of intellectual property protection in life sciences.

3.2.2 Global Case Studies

A. Diamond v. Chakrabarty (U.S.A., 1980)

Jurisdiction: Supreme Court of the United States

Subject: Patentability of a genetically modified organism (GMO)

Summary

Dr. Ananda Mohan Chakrabarty, a microbiologist at General Electric, developed a genetically engineered bacterium capable of breaking down crude oil components—an innovation with applications in oil spill bioremediation. The U.S. Patent Office rejected the patent on grounds that living organisms are not patentable. The Supreme Court overturned this, ruling 5–4 that "anything under the sun that is made by man" is patentable, including living, human-engineered organisms.

Significance

- Set a global precedent for patenting GMOs.
- Marked the beginning of the biotech patent era.

B. Association for Molecular Pathology v. Myriad Genetics (U.S.A., 2013)

Jurisdiction: U.S. Supreme Court

Subject: Patent eligibility of isolated human genes (BRCA1 and BRCA2)

Summary

Myriad Genetics held patents on isolated DNA sequences of BRCA1 and BRCA2 genes associated with breast and ovarian cancer. The patents prevented others from offering diagnostic tests for these genes. The plaintiffs argued that genes are products of nature and should not be patentable.

Ruling

The Court ruled that **naturally occurring DNA** sequences cannot be patented, but **synthetically created complementary DNA (cDNA)** can.

Implications

- Opened access to genetic diagnostics.
- Raised ethical questions about owning human genetic material.

C. Monsanto Co. v. Bowman (U.S.A., 2013)

Jurisdiction: U.S. Supreme Court

Subject: Patent infringement by reuse of genetically modified seeds

Summary

Vernon Bowman, a farmer, purchased patented Roundup Ready soybean seeds from Monsanto and reused harvested seeds for replanting. Monsanto sued for patent infringement.

Ruling

The Court ruled in favor of Monsanto, stating that the **patent exhaustion doctrine** does not permit reproduction of patented seeds without authorization.

Key Learning

- Reinforced patent holder control over self-replicating technologies.
- Impacted seed sovereignty debates worldwide.

D. CRISPR Patent Dispute (U.S.A. and Europe, 2012–present)

Parties: Broad Institute vs. University of California, Berkeley **Subject**: Rights over CRISPR-Cas9 gene-editing technology

Summary

Both parties claimed inventorship of CRISPR-Cas9 for use in eukaryotic cells. The U.S. Patent Trial and Appeal Board awarded rights to Broad Institute (Feng Zhang), while European Patent Office (EPO) invalidated some Broad patents for procedural errors.

Global Relevance

- Multinational implications in biotech innovation.
- Affected investment and licensing in gene editing technologies.

3.2.3 Indian Case Studies

A. Novartis AG v. Union of India (2013)

Court: Supreme Court of India

Subject: Patent eligibility under Section 3(d) of the Indian Patents Act

Summary

Novartis sought a patent for **Glivec (ImatinibMesylate)**, a modified version of an existing anticancer drug. The Indian Patent Office rejected the application, invoking Section 3(d), which bars patents on new forms of known substances unless they demonstrate significant **therapeutic efficacy**.

Ruling

The Supreme Court upheld the rejection, asserting that incremental innovation (so-called "evergreening") is not patentable without enhanced efficacy.

Impact

- Affirmed India's public health-centric IPR stance.
- Ensured availability of affordable generics in India and abroad.

B. Bt Cotton and Monsanto v. Nuziveedu Seeds Ltd. (2019)

Court: Delhi High Court & Supreme Court (ongoing litigation)
Subject: Licensing and patent validity on genetically modified (GM) seeds

Background

Monsanto's Bt cotton technology was licensed to Indian seed companies. Disputes arose over royalty payments and alleged patent invalidity under Indian law.

Key Issue

Are genetically modified seeds patentable in India under Section 3(j) of the Patents Act, which excludes plants and biological processes?

Status

Supreme Court stayed the Delhi High Court's ruling that had invalidated Monsanto's patent. The case is pending final adjudication.

Relevance

- Highlights tension between **IPR and farmers' rights**.
- Raises questions about the **scope of biotechnology patents** in Indian agriculture.

C. Roche v. Cipla (2008)

Court: Delhi High Court

Subject: Enforcement of patent on anti-cancer drug Erlotinib (Tarceva)

Summary

Roche sued Cipla for launching a generic version of Tarceva without a license. Cipla invoked public interest and affordability concerns.

Outcome

The court allowed Cipla to continue production but emphasized the importance of **balancing patent rights with public health**.

Importance

- A precedent for **public interest-based exceptions** in patent enforcement.

D. Yahoo v. Controller of Patents (2021)

Court: Delhi High Court

Subject: Algorithm and bioinformatics-related patentability

Summary

Yahoo's patent application for a bioinformatics algorithm was rejected. The court upheld the decision, stating that computer programs and algorithms are non-patentable under Section 3(k) of the Indian Patents Act.

Takeaway

- Signifies the limitations in patenting **computational biology tools** in India.

3.2.4 Comparative Insights from Case Studies

Aspect	India	Global (U.S./E.U.)
Patent on Life Forms	Restricted under Section 3(j)	Allowed with limitations (e.g., Chakrabarty case)
Diagnostic Methods	Often excluded	Eligible (depending on novelty and utility)
Incremental Innovation	Scrutinized (Section 3(d))	Frequently patented (e.g., evergreening)
Public Interest	Plays a central role (e.g., Novartis, Roche)	Considered but with less emphasis
Gene Editing Tools	Disputed in courts (e.g., CRISPR)	Complex, under evolving IP frameworks

3.2.5 Conclusion

Patent case law in biotechnology reflects the **complex interplay between innovation, legal systems, ethical values, and societal needs**. While global decisions underscore expansive patent coverage, **India's jurisprudence balances innovation with access and public welfare**, offering a model for countries seeking to harmonize IPR with developmental goals. The selected case studies demonstrate how courts mediate the delicate boundaries of what can and should be protected under patent law in life sciences.

3.3 Patentability of Genes, Organisms, and Biotechnological Processes

3.3.1 Introduction

Biotechnology operates at the frontiers of science and law, where **genes, living organisms, and molecular processes** intersect with intellectual property regimes. The **patentability** of biological materials is not only a technical matter governed by statutory rules but also a complex ethical, legal, and social issue. Determining what qualifies as **patent-eligible subject matter** in the realm of biotechnology has prompted debates across jurisdictions. This section explores the **scope, conditions, and controversies** surrounding the patentability of genes, organisms, and biotechnological processes in both international and Indian contexts.

3.3.2 Patentability of Genes

A. Natural vs. Synthetic Genetic Material

A gene, in its **natural state**, is generally **not patentable**, as it is considered a **product of nature**. However, if a gene is **isolated and purified** in a laboratory and serves a **novel, industrial purpose**, it may be patentable in some jurisdictions.

- **United States**: In *Association for Molecular Pathology v. Myriad Genetics (2013)*, the U.S. Supreme Court ruled that **naturally occurring DNA sequences are not patentable**, but **complementary DNA (cDNA)**, which is synthetically constructed, is eligible.
- **European Union**: Under the **Biotech Directive 98/44/EC**, isolated genetic sequences may be patented if their **function is disclosed** and they meet the usual patent criteria.
- **India**: As per **Section 3(c)** of the Indian Patents Act, discoveries of natural substances, including DNA, are not patentable unless they are **substantially modified** or engineered for specific industrial application.

B. Patent Criteria Applied to Genes

To be patentable, genetic inventions must fulfill three basic criteria:

- **Novelty**: The gene sequence must not have been previously disclosed.
- **Inventive step**: The application of the gene should not be obvious to someone skilled in the field.
- **Industrial applicability**: The gene should have a demonstrable use in medicine, agriculture, or industry.

3.3.3 Patentability of Organisms

A. Microorganisms

Microorganisms, especially those genetically modified, have been at the core of patent law debates.

- **Global Precedent**: In *Diamond v. Chakrabarty (1980)*, the U.S. Supreme Court ruled that a **genetically engineered bacterium** capable of degrading oil spills was patentable because it was "a product of human ingenuity."
- **WIPO and TRIPS**: The WTO's **TRIPS Agreement (Article 27.3(b))** allows members to exclude plants and animals from patentability but requires the patenting of microorganisms.
- **India**: Patentable under **Section 3(j)** exception only if microorganisms are **genetically modified** and not naturally occurring.

B. Plants and Animals

- **India**: **Section 3(j)** of the Indian Patents Act prohibits patents on **plants and animals in whole or part**, including seeds and biological processes for their production. Such materials are protected under the **Protection of Plant Varieties and Farmers' Rights Act, 2001 (PPVFR Act)** instead.
- **International Perspective**: Some jurisdictions (e.g., the U.S., Australia) allow patenting of **transgenic animals and plants** if they meet patentability standards.

3.3.4 Patentability of Biotechnological Processes

Biotechnological processes—especially those involving **genetic engineering, recombinant DNA technology, tissue culture, enzyme technology, fermentation**, etc.—are generally patentable, provided they are **novel, involve an inventive step, and are industrially applicable**.

A. Categories of Biotech Processes Eligible for Patents

1. **Recombinant DNA techniques** – e.g., insertion of gene sequences into plasmids.
2. **Gene editing processes** – e.g., CRISPR-Cas9 methodology.
3. **Tissue culture and micropropagation techniques** – when leading to novel outcomes.
4. **Fermentation processes** – for drug production (e.g., insulin, antibiotics).
5. **Protein purification and monoclonal antibody production** – if novel.

B. Exclusions

Certain processes are **non-patentable** due to ethical or legal considerations:

- **Biological processes for the production of plants or animals** (e.g., conventional breeding) are non-patentable in India [Section 3(j)].
- **Processes contrary to public order or morality** (e.g., human cloning, germline modification).

3.3.5 Global and Indian Patent Regime Comparison

Aspect	India	United States / Europe
Patent on Natural Genes	Not allowed (Section 3(c))	Not allowed (post-Myriad case in U.S.; EU requires function disclosure)
Patent on Genetically Modified Organisms	Allowed for microorganisms, not for plants/animals (Section 3(j))	Allowed if novel and non-obvious
Patent on Biotechno-logical Processes	Allowed if not purely biological and if industrially applicable	Broadly allowed, including gene editing, tissue culture, etc.
Ethical Constraints	Strong emphasis on morality and public health	Considered, but less restrictive

3.3.6 Ethical and Societal Considerations

Biotech patentability raises significant concerns:

- **Ownership of life**: Can life forms be "owned" or controlled by corporations?
- **Access to medicines and seeds**: Patents may restrict affordability and availability.
- **Biopiracy**: Patenting of biological resources from developing countries without fair compensation or benefit-sharing.
- **Biodiversity and indigenous rights**: Concerns over commodification of traditional knowledge.

India has addressed these concerns through mechanisms like

- **Traditional Knowledge Digital Library (TKDL)**
- **Biological Diversity Act, 2002**
- **Mandatory disclosure of source and origin of biological material in patent applications**

3.3.7 Conclusion

The **patentability of genes, organisms, and biotechnological processes** remains a **contentious and evolving field**, influenced by scientific advancements, ethical frameworks, international agreements, and national legislation. While patents incentivize innovation, **a balanced approach** is essential to safeguard **public interest, bioethics, environmental sustainability**, and **access to essential technologies**. A harmonized yet context-sensitive patent regime is indispensable in fostering responsible biotechnology development.

4

Patent Application, Procedure and Treaties

Section 4.1 Patent Application Process in India

4.1.1 Introduction

India's patent system is governed by the **Patents Act, 1970**, as amended by the **Patents (Amendment) Acts of 1999, 2002, and 2005**, and is administered by the **Office of the Controller General of Patents, Designs, and Trade Marks (CGPDTM)**. The Act aligns with the **Trade-Related Aspects of Intellectual Property Rights (TRIPS)** Agreement, ensuring compliance with global standards. For inventions in **modern biotechnology**, the patent application process in India involves a rigorous examination of novelty, inventive step, industrial applicability, and statutory exclusions, particularly under Sections **3(b), 3(c), 3(d), and 3(j)**.

4.1.2 Who Can Apply?

Under **Section 6** of the Patents Act, a patent application in India can be filed by:

- The **true and first inventor**,
- An **assignee** of the inventor,
- A **legal representative** of a deceased inventor.

In the case of collaborative research or industrial research programs, both academic institutions and corporate entities can apply jointly.

4.1.3 Types of Patent Applications in India

Type	**Description**
Provisional Application	Filed to secure a priority date; must be followed by a complete specification within 12 months.
Complete Application	Contains full disclosure, including claims and abstract; eligible for examination.
Convention Application	Filed within 12 months of filing in a convention country.
PCT National Phase	Filed in India based on an international application under the Patent Cooperation Treaty (PCT).

4.1.4 Step-by-Step Patent Application Process in India

1. Patentability Assessment

- Preliminary evaluation is conducted to ensure the invention is **novel**, **inventive**, and **industrially applicable**.
- **Patent agents or legal professionals** often assist in drafting claims and specifications in biotech inventions, which require technical precision.

2. Patent Search

- Conducted using databases like **IPO Patent Search System (InPASS)**, **WIPO Patentscope**, or **Espacenet** to avoid duplication and prior art conflicts.

3. Filing the Application

- Application is submitted **online** (via IP India portal) or **physically** to one of the four Patent Offices in India (Delhi, Mumbai, Chennai, Kolkata).
- **Form 1**: Application for grant of patent
- **Form 2**: Provisional or complete specification
- **Form 3**: Statement and undertaking under Section 8
- **Form 5**: Declaration of inventorship
- **Form 18**: Request for examination
- **Form 26**: If a patent agent is involved

4. Publication

- Application is published in the **Patent Journal** after **18 months** from the filing or priority date.
- Early publication may be requested through **Form 9** for strategic purposes.

5. Request for Examination (RFE)

- **Must be filed within 48 months** from the priority date using **Form 18**.
- Failing to file RFE leads to abandonment of the application.

6. Examination

- The application is assigned to a **Patent Examiner** who issues a **First Examination Report (FER)** based on:
  - Prior art search
  - Patentability under Sections 3 and 4
  - Clarity and sufficiency of disclosure

- The applicant must respond to FER within **6 months**, extendable by **3 months**.

7. Pre-Grant Opposition

- Allowed under **Section 25(1)** by any person after publication but before grant.
- Grounds include lack of novelty, prior claims, wrongful acquisition, or violation of public morality.

8. Grant of Patent

- If all objections are addressed satisfactorily, the Controller grants the patent, which is published in the Patent Journal.
- The **patent term is 20 years** from the filing date.

9. Post-Grant Opposition

- Can be filed under **Section 25(2)** within **12 months** from the grant date by an interested party.

10. Maintenance and Renewal

- Patents must be renewed annually by paying renewal fees from the **3rd year onwards**.
- Failure to pay results in patent lapse.

4.1.5 Special Considerations for Biotechnology Inventions

Biotechnology-related applications often attract scrutiny under specific exclusions:

Section	Provision
3 (b)	Inventions contrary to public order or morality (e.g., human cloning).
3 (c)	Discovery of living/non-living substances occurring in nature (e.g., natural genes).
3 (d)	New forms of known substances without increased efficacy (e.g., polymorphs).
3 (j)	Plants, animals (in whole or part), seeds, and biological processes.

To strengthen a biotechnology application:

- Clear **technical disclosure**, including sequences, pathways, or structural diagrams, must be provided.
- Novelty must be demonstrated with **comparative data** or performance metrics.
- For **biological materials**, deposition in a **recognized international depository (e.g., MTCC or ATCC)** is recommended.

4.1.6 Role of Traditional Knowledge and Biodiversity Laws

India mandates **disclosure of source and geographical origin** of biological material (under Section 10(d)(ii) of the Patents Rules). Additionally:

- Applicants must **seek approval from the National Biodiversity Authority (NBA)** under the **Biological Diversity Act, 2002**, before applying for patents on biological resources or associated traditional knowledge.
- Failure to comply may result in **rejection or revocation** of the patent.

4.1.7 Timelines and Strategic Filing Tips

Activity	Timeframe
Filing Provisional + Complete	Complete within 12 months
Publication	18 months from priority date
Request for Examination	Within 48 months
First Examination Report Response	6 months + 3 months extension
Opposition (Post-Grant)	Within 12 months from grant date

Strategic Tips

- Use **provisional filing** to secure early priority for cutting-edge biotechnological inventions.
- Consider **international filing via PCT** within 12 months if global protection is desired.
- Maintain **clear laboratory records** and invention disclosures to support patent claims.

4.1.8 Conclusion

The Indian patent application process is robust, transparent, and aligned with global norms while being sensitive to the country's biodiversity, ethical, and socio-economic concerns. For modern biotechnology innovations, it offers both opportunity and complexity. Navigating it requires **scientific clarity**, **legal diligence**, and **regulatory compliance**, particularly regarding exclusions related to biological materials, ethical considerations, and biodiversity laws. The Indian system continues to evolve to strike a balance between promoting innovation and safeguarding public interest, especially in life sciences and biotechnology.

4.2 International Patent Filing – The Patent Cooperation Treaty (PCT)

4.2.1 Introduction to the PCT System

The **Patent Cooperation Treaty (PCT)** is a multilateral treaty administered by the **World Intellectual Property Organization (WIPO)**, established in **1970** and entered into force in **1978**. It provides a **unified, streamlined**

mechanism for filing patent applications in multiple countries through a **single international application**, thereby simplifying the process for inventors and organizations seeking **global patent protection**.

India became a **signatory to the PCT in 1998**, and since then, Indian applicants have leveraged this route to extend their patent rights beyond national borders—particularly important in the biotechnology sector, where innovations often have cross-border commercial and therapeutic significance.

4.2.2 Objectives and Benefits of the PCT

The PCT does **not grant international patents**, but it facilitates:

- Filing a **single application** in **one language**, with **one set of fees**, to seek protection in **over 150 contracting states**.
- Conducting an **International Search** and optionally an **International Preliminary Examination** to evaluate patentability before entering national phases.
- **Postponement of costs** and strategic decisions regarding countries of protection for up to **30/31 months** from the priority date.

For biotechnology inventions, which often require substantial validation and international collaboration, the PCT system offers critical advantages in terms of:

- **Time for experimental validation**
- **Freedom to operate analyses**
- **Investor engagement prior to national entry**

4.2.3 Key Terminologies in PCT Process

Term	**Explanation**
Receiving Office (RO)	Office where the PCT application is filed (e.g., IPO India).
International Searching Authority (ISA)	Authority that conducts prior art search and issues search report.
International Preliminary Examining Authority (IPEA)	Optional body that conducts a detailed patentability examination.
International Bureau (IB)	Administered by WIPO; processes and publishes international applications.

4.2.4 Steps in the PCT Filing Process

Step 1: Filing the International Application

- Applicant files an **international application** with:
  - **RO** (e.g., Indian Patent Office) or directly with **WIPO**
  - In **English or Hindi**

    - Claiming **priority** from an earlier national application (within 12 months under the Paris Convention)
- Application includes
    - **Request Form (PCT/RO/101)**
    - **Description**, **claims**, **abstract**, **drawings** (if any)
    - Prescribed **international filing fee**, **search fee**, and **transmittal fee**

Step 2: International Search

- Conducted by an **ISA** (India uses ISA/IN or other recognized ISAs like ISA/EP, ISA/US)
- Results in
    - **International Search Report (ISR)** – lists prior art documents
    - **Written Opinion (WO)** – gives preliminary view on patentability

Step 3: Publication by WIPO

- Application is published **after 18 months** from the priority date in the **WIPO PatentScope database**
- Includes ISR, making the application and its search public

Step 4: International Preliminary Examination (Optional)

- Requested using **Form PCT/IPEA/401**
- Conducted by an IPEA to assess inventive step, novelty, and industrial applicability
- Results in **International Preliminary Report on Patentability (IPRP Chapter II)**

Step 5: Entry into National/Regional Phases

- Must be done within:
    - **30 months** (for most countries, including India, USA, China)
    - **31 months** (for a few others like Japan, Canada)
- Applicant files in individual countries or regional offices (like EPO, ARIPO, etc.)
- Must comply with local laws, translations, and fee requirements

4.2.5 Structure and Timeline of PCT Process

Stage	Timeframe (from priority date)
International filing	Within 12 months
International search & WO	~16 months
International publication	18 months
Optional preliminary exam	By 22 months
National phase entry	By 30/31 months

4.2.6 Role of PCT in Biotechnology

The PCT mechanism is especially important in **biotechnological innovation**, where:

- The global commercialization of vaccines, diagnostics, and GMOs requires **broad territorial rights**.
- **Early-stage inventions** may need **further validation**, making the extended timeline of PCT attractive.
- PCT search and preliminary examination offer **predictive insights** into how patent offices may treat the invention.
- **Depositing biological materials** (e.g., microbial strains) under the **Budapest Treaty** can be coordinated with PCT filings.

4.2.7 India's Participation in the PCT Framework

- India acts as
  - **Receiving Office (RO/IN)**
  - **International Searching Authority (ISA/IN)** at the Indian Patent Office (Delhi)
  - **International Preliminary Examining Authority (IPEA/IN)**
- Indian researchers, start-ups, and academic institutions can **directly file PCT applications** or via national route.

Example

A novel CRISPR-Cas9 delivery vector developed by an Indian biotech start-up can be first filed as an Indian patent, and within 12 months, filed as a PCT application to secure global rights.

4.2.8 Legal and Strategic Implications

- Filing a PCT application does **not guarantee** patent grant in any country; **national laws** prevail during the national phase.
- Strategic use of PCT can

- Postpone high costs of global filing
- Improve **licensing potential** with validated prior art reports
- Strengthen **technology valuation**

4.2.9 Challenges and Limitations

Challenge	**Explanation**
High cost of national phase entry	Filing, translation, and attorney fees in multiple jurisdictions
Patentability variations	Different nations may have differing standards, especially for biotech inventions (e.g., Europe vs. US on gene patenting)
Delays in national processing	Especially in jurisdictions with long examination backlogs

4.2.10 Conclusion

The PCT system revolutionizes international patent filing by offering a **consolidated and strategic pathway** for global protection of inventions. For biotechnology innovations, the PCT process enhances global visibility, enables informed business decisions, and delays cost-intensive steps until the invention matures technically and commercially. India's active participation as an ISA and IPEA offers domestic innovators a globally compliant yet localized support system for extending the reach of their intellectual creations.

Section 4.3 Documentation and Legal Requirements

4.3.1 Introduction

In the realm of biotechnology patenting, the **integrity, accuracy, and completeness of documentation** are paramount to securing and maintaining legal rights over an invention. Patent documentation is not merely a formality—it embodies the **technical disclosure**, **legal justification**, and **claim boundaries** of an innovation. Understanding the **statutory requirements**, **formatting standards**, and **submission protocols** for both national and international filings is essential for inventors, legal professionals, and biotechnology entrepreneurs alike.

4.3.2 Essential Documents for Filing a Patent Application in India

In accordance with the **Indian Patent Act, 1970 (as amended)** and **Patent Rules, 2003**, the following documents are mandatory for a patent application:

Document Name	Purpose
Form 1: Application for Grant of Patent	General information: applicant, inventor, type of application, declaration.
Form 2: Provisional/Complete Specification	Technical disclosure of the invention. Provisional can be filed initially.

Form 3: Statement and Undertaking	Information regarding foreign filings. Must be updated periodically.
Form 5: Declaration as to Inventorship	Declares the true and first inventor(s).
Form 26: Power of Attorney	Required if filed through a patent agent.
Form 18: Request for Examination (RFE)	Must be filed within 48 months from priority date.
Form 27: Working of Patent Statement	Filed post-grant to indicate commercial use in India.

Note: Forms 4, 6, 7–25 are situation-specific and may be required for divisional applications, post-grant proceedings, opposition, corrections, or amendments.

4.3.3 Legal Requirements for a Valid Patent Application

To ensure that the application is **legally admissible and enforceable**, the following statutory requirements must be satisfied:

1. Patentable Subject Matter

- The invention must fall within **patent-eligible categories** under Section 3 and 4 of the Indian Patents Act.
- In biotechnology, exclusions include:
  - Methods of agriculture or horticulture
  - Diagnostic, therapeutic, and surgical methods for humans and animals
  - Biological materials per se, unless isolated and industrially applicable

2. Novelty

- The invention must be **new** and not anticipated by prior art.

3. Inventive Step (Non-obviousness)

- The invention must involve **technical advancement** or **economic significance**, and not be obvious to a person skilled in the art.

4. Industrial Applicability

- The invention must be **capable of being made or used** in an industry, particularly relevant to biotech processes, diagnostic kits, or microbial production.

5. Full and Enabling Disclosure

- The complete specification must disclose:
  - The best method of performing the invention
  - Clear, sufficient, and reproducible instructions
  - Sequence listings (in case of nucleic acid/protein inventions) as per WIPO ST.25

6. Clarity and Unity of Invention

- Claims must be **clear, concise**, and **supported** by the description.
- The application must relate to **one invention** or a **group of linked inventions** (unity of invention).

4.3.4 Supporting Documentation in Biotechnology Patents

In addition to the formal patent documents, biotechnology patent applications often require the following:

A. Sequence Listings

- For inventions involving genes, proteins, or nucleotides, a **separate sequence listing** (in electronic and paper format) is required.
- Must adhere to **WIPO ST.25** or newer **ST.26 XML-based standard**.

B. Deposit Certificates (for Biological Materials)

- If the invention involves a **microorganism**, it must be deposited in an **International Depository Authority (IDA)** under the **Budapest Treaty**.
- India recognizes **MTCC (IMTECH, Chandigarh)** and **NCIM (Pune)** as IDAs.

C. Prior Art References

- Disclosure of known related inventions or literature to comply with **Form 3** and to assist in prior art analysis.

D. Inventor Consent Forms

- Declarations ensuring that inventors have willingly assigned or co-applied for the patent.

4.3.5 International Filing: Additional Legal Documentation

When filing through the **PCT route**, the following documents are essential:

Document	Purpose
Request Form (PCT/RO/101)	Standard form for PCT application
Description, Claims, Abstract	Core content of the application
Sequence Listings	Required if nucleotide/amino acid sequences are disclosed
International Search Report (ISR)	Prepared by ISA
Written Opinion (WO)	Preliminary analysis of patentability
Priority Document	Copy of the earlier application under the Paris Convention
Declaration of Inventorship	Identifying true inventors

For smooth PCT processing, all documents must comply with **WIPO filing standards**, **translation requirements**, and **formatting rules** (e.g., typed in Unicode-compliant font, A4 paper, with line numbering).

4.3.6 Confidentiality and Disclosure Norms

- **Section 8** of the Indian Patents Act mandates the disclosure of corresponding foreign applications.
- Pre-filing public disclosures (e.g., journal articles, conferences) can **destroy novelty** unless covered under **grace period provisions** (applicable only under specific jurisdictions).
- Maintaining **lab notebooks**, **signed records**, and **experimental proof** is critical for defending inventorship and priority in contentious cases.

4.3.7 Digital and Electronic Filing Requirements

- India allows **e-filing** through the **Comprehensive e-Filing Services of IPO**, which includes:
  - Digital signature verification
  - Auto-acknowledgment and document tracking
- Internationally, the **ePCT system** by WIPO offers a secure portal for uploading, amending, and managing international applications.
- Mandatory formats:
  - PDF for specifications
  - XML for sequence data
  - DOC/DOCX for editable forms (pre-upload)

4.3.8 Common Errors and Legal Pitfalls

Common Error	**Consequence**
Incomplete specification	Leads to rejection or post-grant revocation
Non-disclosure of foreign filings	May result in abandonment or penalties under Section 8
Incorrect inventorship declaration	Legal disputes over ownership
Omitted sequence listing	Grounds for non-compliance in biotech applications
Missed deadlines for RFE/Form-18	Application may be treated as withdrawn

4.3.9 Post-Grant Legal Requirements

Once a patent is granted, legal obligations continue:

- **Payment of annuity fees** (yearly)
- **Submission of Form 27** (working of patent) annually
- **Maintenance of patent records** in case of technology transfer or licensing

- **Opposition compliance** – defending the patent in case of post-grant opposition under Section 25(2)

4.3.10 Conclusion

The patent application process in biotechnology is not just a technical exercise—it is a **legally intricate process** governed by domestic laws, international treaties, and sector-specific disclosure norms. A well-documented application with clear claims, detailed specifications, and statutory compliance enhances the likelihood of grant and ensures that the rights can be **legally defended**, **commercially exploited**, and **internationally harmonized**. Careful adherence to documentation and legal requirements protects the innovation pipeline and supports the sustainable development of biotechnology in India and globally.

4.4 Timeframes and Patent Enforcement

4.4.1 Introduction

The lifecycle of a patent is governed by a series of critical **statutory timeframes** that define the procedural validity, commercial strength, and enforceability of patent rights. For biotechnology innovations, where the **R&D cycle is long**, and **market stakes are high**, a clear understanding of **application deadlines**, **examination timelines**, **renewals**, and **enforcement mechanisms** is vital. Effective patent enforcement further ensures that rights conferred through patent grants are **protected against infringement**, both in India and globally.

4.4.2 Statutory Timeframes in the Indian Patent System

The **Indian Patents Act, 1970 (as amended)** and **Patent Rules, 2003** define the **sequential timeframes** that every applicant must adhere to:

Stage	**Timeframe**	**Consequences of Delay**
Filing of **Provisional Specification**	Optional initial filing	Must file complete specification within **12 months**
Filing of **Complete Specification**	Within **12 months** of provisional filing	Application deemed **abandoned** if missed
Filing of **Request for Examination (Form 18)**	Within **48 months** from date of priority	Application **withdrawn** if not filed
Publication of application	Automatically after **18 months** from priority	Can be **requested earlier** via Form 9
First Examination Report (FER) issuance	Generally within **6–12 months** post RFE	FER raises objections to be addressed by applicant
Response to **FER**	Within **6 months**, extendable by **3 months**	Failure results in **application deemed abandoned**
Filing of **Pre-grant opposition**	Any time **before grant**	Patent Office may reject based on opposition

Filing of **Post-grant opposition**	Within **12 months** of patent grant	Adjudicated by **Opposition Board**
Filing of **Working of Patent Statement (Form 27)**	**Annually**, post-grant	Non-compliance can lead to **revocation**
Renewal fees (Annuity payments)	From **3rd year onwards**, till **20th year**	Lapse of patent for non-payment of annuity

4.4.3 Timeframes Under the PCT System

The **Patent Cooperation Treaty (PCT)** offers a unified application route with its own time-bound procedures:

Procedure	**Deadline**
Filing of International Application	Within **12 months** of earliest priority claim
International Search Report (ISR)	Issued within **16 months** from priority
International Publication	Occurs after **18 months** from priority
Filing **Demand for International Preliminary Examination** (Optional)	Within **22 months** from priority
Entry into National Phase	Within **30 or 31 months** (India allows 31 months)

Biotechnology applicants often use the **PCT route** to gain **early search reports** and **extended time** for market evaluation before committing to national phase entry.

4.4.4 Patent Term and Expiry

In India, the **term of a patent** is uniformly fixed at **20 years** from the **date of filing** of the patent application, regardless of whether it was filed with a provisional or complete specification.

Biotechnology-specific aspects

- For PCT applications, the 20-year term is counted from the **International Filing Date**.
- There is **no provision for patent term extension** in India, unlike in some jurisdictions like the US (where patent term adjustment for regulatory delays is available).

4.4.5 Grace Period and Restoration

- India does **not provide a general grace period** for pre-filing disclosures by the inventor. Any public disclosure before filing may destroy novelty.
- However, **Section 72 and 60** of the Indian Patent Act allows for **restoration** of lapsed patents due to non-payment of renewal fees:
    - Application for restoration must be made within **18 months** from the date of lapse.

- Restoration is subject to **proof of unintentional delay** and payment of fee.

4.4.6 Patent Enforcement Mechanisms in India

Patent enforcement is the process of **legally asserting patent rights** against unauthorized users (infringers). In India, enforcement is **civil in nature** and governed by the **Indian Patents Act, 1970** and **Civil Procedure Code**.

Legal Forums for Enforcement

- **District Courts** (original jurisdiction)
- **High Courts** (for revocation or appeals)
- **Intellectual Property Division (IPD)** of High Courts (e.g., Delhi)

4.4.7 Types of Patent Infringement

Type	Description
Direct Infringement	Unauthorized making, using, selling, or importing patented item
Contributory Infringement	Supplying components knowing they will be used to infringe
Literal Infringement	Product/process directly violates claims as written
Doctrine of Equivalents	Slight modifications still infringe due to equivalent effect

4.4.8 Remedies Against Infringement

An aggrieved patentee can seek the following legal remedies:

1. Injunction

- **Temporary injunction**: to prevent immediate harm
- **Permanent injunction**: granted post-trial to bar continued use

2. Damages or Account of Profits

- Monetary compensation for losses due to infringement
- In some cases, profits made by infringer are recoverable

3. Delivery Up or Seizure

- Court can order infringing products to be seized or destroyed

4. Declaration of Validity

- In contested cases, court may affirm patent's validity

4.4.9 Burden of Proof in Biotech Process Patents

Section 104A of the Indian Patents Act addresses **burden of proof reversal** for **process patents**, especially in biotechnology:

If a patented process leads to a product, and the patentee proves that the product is identical to the patented one, then the defendant must prove that a different process was used.

This is crucial in **microbial fermentation**, **recombinant protein expression**, and **bioprocess engineering** where detection of infringement is difficult.

4.4.10 Customs and Border Enforcement

Under the **Intellectual Property Rights (Imported Goods) Enforcement Rules, 2007**, Indian customs authorities can **intercept and seize infringing goods** at borders, if:

- The patent is registered with the **Customs IPR Recordation Portal**
- Sufficient documentary evidence is provided
- Biotech items like patented seeds, diagnostics, or therapeutics can be protected from **parallel imports** and **counterfeit entry**

4.4.11 International Enforcement Considerations

Patents are **territorial rights**. Enforcement must be pursued **individually** in each country where rights are granted. Global patent holders must:

- **File for protection** in countries of commercial interest
- Use treaties like **TRIPS**, **WIPO Arbitration and Mediation**, and **Bilateral Treaties** for enforcement support
- In case of cross-border biotech disputes, turn to **jurisdiction-specific patent courts** (e.g., UPC in EU, CAFC in US)

4.4.12 Conclusion

A strong biotechnology patent is only as effective as its timely prosecution and robust enforcement. Adherence to statutory **deadlines**, **renewals**, and **opposition periods** ensures procedural strength, while knowledge of **legal remedies**, **proof dynamics**, and **international enforcement** arms patent holders against infringers. In the highly competitive world of biotechnology, mastering the **time-bound enforcement landscape** is indispensable for safeguarding innovation and market exclusivity.

5

Plant Variety Protection and Copyright in Biotechnology

5.1 Plant Breeder's Rights (PBR) and Protection of Plant Varieties and Farmers' Rights (PPV&FR) Act

5.1.1 Introduction

In the context of biotechnology and agricultural innovation, **Plant Breeder's Rights (PBRs)** offer exclusive intellectual property protection to individuals or organizations that breed, develop, or discover new plant varieties. These rights ensure **economic incentives** for plant breeders while promoting **innovation in seed development** and **agricultural productivity**. In India, the unique and inclusive **Protection of Plant Varieties and Farmers' Rights Act (PPV&FR), 2001**, balances breeders' rights with **recognition of farmers' contributions**, making it one of the most progressive legislations in the world.

5.1.2 Concept of Plant Breeder's Rights (PBR)

Plant Breeder's Rights refer to the **exclusive rights** granted to plant breeders to **produce, sell, market, distribute, import or export** their registered plant varieties. These rights are **sui generis**, meaning they are a special form of IP distinct from patents, developed to cater specifically to plant varieties.

Key Objectives of PBR

- Encourage investment in plant breeding and biotechnology.
- Ensure access to improved crop varieties.
- Prevent unauthorized commercial exploitation of novel varieties.
- Facilitate biodiversity conservation through formal registration systems.

5.1.3 International Framework for PBR: UPOV Convention

The **International Union for the Protection of New Varieties of Plants (UPOV)** provides a global framework for plant variety protection. UPOV was established by the **UPOV Convention (1961)**, with subsequent revisions in 1972, 1978, and 1991.

UPOV 1991 Features

- Extended breeder's rights to harvested materials and essentially derived varieties (EDVs).
- Allowed greater flexibility for national PVP systems.
- Restricted the traditional rights of farmers to save and reuse seeds.

India is **not a member of UPOV** due to concerns over farmers' rights. Instead, it has adopted a **national sui generis system** under the PPV&FR Act, 2001.

5.1.4 The Protection of Plant Varieties and Farmers' Rights (PPV&FR) Act, 2001

India's PPV&FR Act is a comprehensive legislation that **simultaneously protects the rights of breeders and farmers**, recognizing the latter as custodians of traditional knowledge.

Salient Features of the PPV&FR Act

Component	Description
Novelty Requirement	Variety must not have been sold in India >1 year, or internationally >6 years.
Distinctiveness	Must be clearly distinguishable from existing varieties.
Uniformity	Should be sufficiently uniform in key traits.
Stability	Traits must remain unchanged over successive generations.
Registration Validity	15 years for trees and vines; 9 years for others (renewable).
Authority	PPV&FR Authority under the Ministry of Agriculture & Farmers' Welfare.

5.1.5 Categories of Registrable Varieties

Under the Act, the following types of plant varieties can be registered:

- **New Variety** – Not commercialized before filing.
- **Extant Variety** – Already in cultivation or in public domain.
- **Farmers' Variety** – Traditionally cultivated and evolved by farmers.
- **Essentially Derived Variety (EDV)** – Developed from a protected variety retaining essential traits.

5.1.6 Rights Granted Under PPV&FR

To Breeders

- Exclusive rights to produce, sell, market, distribute, import, or export the variety.
- Rights extend to EDVs and harvested materials.

To Farmers

- Right to save, use, sow, re-sow, exchange, share or sell farm produce (except branded seeds).
- Right to register and protect traditional varieties.
- Right to compensation if a registered variety fails to perform as claimed.

To Researchers

- Use of protected varieties for research and for creating new varieties is permitted.

5.1.7 Benefit Sharing and Compensation

India's PPV&FR Act mandates **benefit-sharing mechanisms** for communities or individuals who have contributed genetic resources used in developing new varieties. A **National Gene Fund** has been established to:

- Provide monetary compensation to farmers.
- Fund conservation and sustainable use of agro-biodiversity.
- Support public awareness and farmer training.

5.1.8 Compulsory Licensing and Public Interest

The Act empowers the Authority to grant **compulsory licenses** after 3 years from registration if:

- The breeder fails to meet the demand for seeds.
- The price of seeds is unreasonable.
- Public interest in food security is at stake.

This provision is especially significant in biotechnology where monopolization of genetically modified seeds could threaten **seed sovereignty**.

5.1.9 Legal Protection and Remedies

Breeders and farmers can seek legal remedies in cases of:

- **Infringement** of registered rights.
- **Misrepresentation** during registration.
- **Unauthorized commercialization** of registered varieties.

Penalties include **fines**, **imprisonment**, and **revocation of registration** in severe cases.

5.1.10 Biotechnology and PBR Interface

In the age of genetic modification and molecular breeding, biotechnology plays a key role in:

- **Marker-assisted selection** for trait-specific breeding.
- Development of **transgenic crops** with novel traits.
- **Gene pyramiding** and hybrid technology.

While biotechnological inventions may be protected under **patents**, the resultant plant varieties must comply with **PBR registration norms**, ensuring harmony between innovation and biodiversity protection.

5.1.11 Challenges and Future Directions

Challenges	Implications
Limited awareness among farmers	Under-registration of traditional varieties and limited benefit sharing
Complexity of DUS testing	Delays in registration and lack of access to technology by grassroots breeders
Conflicts with Patent Law (Section 3 (j))	Restriction on patenting plants but unclear scope of biotechnological claims
Private sector monopolies	Need for stronger enforcement and equitable seed pricing policies

Future reforms must focus on **digitizing DUS procedures**, promoting **community seed banks**, and encouraging **public-private partnerships** in crop improvement under **ethically governed IP frameworks**.

5.1.12 Conclusion

The **PPV&FR Act of India** represents a globally significant model of **inclusive and equitable IP protection** in the realm of agriculture and biotechnology. By simultaneously recognizing **breeders**, **farmers**, and **researchers**, the Act builds a rights-based ecosystem that **encourages innovation**, **protects traditional knowledge**, and ensures **seed sovereignty**. In the broader canvas of biotechnology IPR, **Plant Breeder's Rights** under this Act remain a vital tool for sustainable innovation, socio-economic equity, and national food security.

Section 5.2 Copyright Issues in Biotech Innovation

5.2.1 Introduction

While **copyright** is often associated with literary, artistic, or musical works, its significance in **biotechnology** has expanded with the digitization of biological data, software-assisted innovations, and the rise of **bioinformatics and computational biology**. In the biotechnology domain, **copyright law** protects the **expression** of original ideas in tangible forms—such as **databases**, **algorithms**, **software**, **lab manuals**, **molecular structure illustrations**, and **training materials**—though **not the underlying scientific ideas or inventions themselves**, which are covered under **patents**.

As biotechnology becomes increasingly **data-driven**, copyright emerges as an essential layer of intellectual property protection, especially in safeguarding **digital knowledge assets** and **proprietary tools**.

5.2.2 Scope of Copyright in Biotechnology

Copyright in biotechnology applies to **creative and original expressions** that are **fixed in a tangible medium**, including but not limited to:

- **Bioinformatics Software Tools** (e.g., genome browsers, sequence alignment tools)
- **Gene and Protein Sequence Databases** (e.g., curated databases like UniProt, GenBank)
- **Visualizations of Molecular Structures**
- **Research Reports and Technical Documentation**
- **Training Manuals, Laboratory Protocols, SOPs**
- **Animations, Diagrams, and Illustrations used in Communication of Science**

Example: A software program written for CRISPR-Cas9 guide RNA design can be copyrighted as a software artifact, even though the genome-editing technique itself is protected by patents.

5.2.3 Copyright vs Patent: A Distinction in Biotech IP

Criteria	**Patent**	**Copyright**
Protects	Inventions, processes, compositions, organisms	Expression of ideas, not the idea itself
Coverage in Biotech	GMOs, gene sequences, bioprocesses	Software, databases, graphical content
Term of Protection	20 years from filing date	Life of author + 60 years (India)
Requirement	Novelty, non-obviousness, industrial applicability	Originality and fixation in a tangible form
Registration	Mandatory (for enforceability)	Optional, but recommended for legal recourse

5.2.4 Indian Legal Framework for Copyright

India's copyright law is governed by the **Copyright Act, 1957**, as amended from time to time, especially by the **Copyright (Amendment) Act, 2012**. The law offers protection for original works in the following categories relevant to biotechnology:

- **Literary Works**: Lab protocols, technical writings, user manuals
- **Computer Programs**: Bioinformatics software, simulations

- **Databases**: Annotated sequence repositories, curated datasets
- **Artistic Works**: Graphical representations of cellular processes or molecular models

In India, **copyright arises automatically** upon creation and fixation of the work; however, **registration serves as prima facie evidence in court proceedings**.

5.2.5 Copyrightable Elements in Biotechnology

a) Software and Computational Tools

Custom-built software tools used for:

- DNA/RNA sequence analysis
- 3D modeling of protein structures
- Metabolic pathway simulations
- Image analysis for microscopy data

Example: A GUI-based interface designed for visualizing protein folding models may qualify for copyright protection.

b) Biological Databases

Databases that involve **curation, annotation, and structuring** of biological information, such as:

- SNP databases
- Gene ontology datasets
- Custom plasmid maps or genome maps

Although raw biological sequences are not copyrightable, **the curated arrangement or annotations** can be.

c) Educational and Research Content

- Instructional materials, PowerPoint presentations, and e-learning modules used in biotechnology education.
- Lab protocols and methods developed and illustrated in a unique format.

5.2.6 International Treaties Relevant to Copyright in Biotech

- **Berne Convention (1886)**: India is a signatory; ensures protection of copyright across member states without the need for formal registration.
- **TRIPS Agreement (WTO)**: Mandates minimum standards of protection for software and databases, which are especially relevant in biotechnology.

- **WIPO Copyright Treaty (1996)**: Emphasizes protection of digital content and software—key in bioinformatics.

5.2.7 Challenges in Copyright Protection of Biotech Works

Issue	Implication
Originality in Scientific Writing	Repetitive protocols and standardized methodologies may lack originality
Overlap with Patent or Trade Secret	Complex ownership questions for hybrid works (e.g., patented algorithms in copyrighted software)
Access vs. Ownership Dilemma	Open-source bioinformatics tools challenge traditional copyright models
Digital Piracy	Widespread digital distribution makes unauthorized use hard to detect
Collaborative Authorship and Attribution	Multi-author research leads to unclear copyright ownership

5.2.8 Copyright and Open Science

A significant segment of biotechnology relies on **open access to data and tools**—especially in genomics and systems biology. Yet, the developers of such tools also deserve acknowledgment and protection.

- **Creative Commons Licenses (CC-BY, CC-BY-SA, etc.)** are often applied to open-access datasets and tools.
- **GNU General Public License (GPL)** or **MIT License** is common for bioinformatics software.

Balanced copyright policies enable **knowledge sharing while ensuring recognition** to the creators.

5.2.9 Best Practices in Copyright Management

- **Documentation of authorship and date of creation**
- **Proper attribution and citation standards**
- **Use of copyright notices and licensing information**
- **Incorporation of Digital Object Identifiers (DOIs)** for software and datasets
- **Institutional IP policies** to define rights of employees and students

5.2.10 Conclusion

Copyright plays a **crucial yet often overlooked role** in the domain of biotechnology. As **bioinformatics**, **data-driven research**, and **digital innovation** take center stage, the need for clear and enforceable copyright protections grows stronger. From proprietary software to curated gene databases and instructional design, a well-implemented copyright regime fosters **creative expression, data integrity**, and **academic recognition** in modern biotechnology.

By understanding and employing copyright strategically, biotech researchers, institutions, and enterprises can **safeguard their intellectual outputs, encourage responsible sharing**, and **promote innovation** in the evolving knowledge economy.

5.3 Distinctions between PVP, Patents, and Copyrights

5.3.1 Introduction

In the realm of **intellectual property rights (IPR)**, three pivotal regimes—**Plant Variety Protection (PVP)**, **Patents**, and **Copyrights**—serve distinct legal and practical functions. While they all aim to safeguard intellectual creativity and innovation, their **scope, subject matter, legal framework, duration of protection, and enforcement mechanisms** differ significantly.

Understanding the **distinctions** between these forms of IPR is crucial for biotechnology professionals engaged in **plant breeding, molecular innovation, software development, data curation**, and **scientific publishing**. This section systematically outlines and compares these IPR types, particularly in the context of biotechnology and agricultural innovation.

5.3.2 Overview of the IPR Types

IPR Type	**Scope of Protection**	**Relevant to Biotechnology**
Plant Variety Protection (PVP)	Protects new plant varieties bred with distinct, uniform, and stable traits	Plant breeders, seed developers, genetic resource conservation
Patent	Protects inventions (e.g., genes, processes, devices) that are novel, non-obvious, and industrially applicable	Transgenic organisms, biotechnological processes, diagnostics
Copyright	Protects original expressions in fixed media (e.g., texts, images, software, databases)	Bioinformatics software, lab protocols, digital content

5.3.3 Legal Basis and Indian Context

Aspect	**PVP**	**Patent**	**Copyright**
Governing Law (India)	*The PPV&FR Act, 2001*	*The Patents Act, 1970 (Amended 2005)*	*The Copyright Act, 1957 (Amended 2012)*
Administering Body	Protection of Plant Varieties and Farmers' Rights Authority	Controller General of Patents, Designs & Trade Marks	Copyright Office, Ministry of Commerce & Industry
International Treaty	UPOV Convention (India not a member)	TRIPS Agreement, PCT (India is a signatory)	Berne Convention, WIPO Copyright Treaty

5.3.4 Subject Matter and Eligibility

Criteria	PVP	Patent	Copyright
What it protects	New plant varieties	Inventions, processes, compositions, organisms	Literary and artistic works, software, data-bases
Eligibility	Novelty, Distinctness, Uniformity, Stability	Novelty, Inventive step, Industrial applicability	Originality and fixation in a tangible medium
Exclusions	Wild species, traditional varieties (unmodified)	Laws of nature, abstract ideas, plants (in India)	Scientific principles, ideas, facts

5.3.5 Duration and Rights Conferred

Aspect	PVP	Patent	Copyright
Term of Protection	15 years (extensible to 18) for extant varieties; 20–25 years for new varieties	20 years from date of filing	Life of author + 60 years (India)
Rights Granted	Exclusive rights to produce, sell, market, distribute, and license the variety	Exclusive rights to make, use, sell, or license	Exclusive rights to reproduce, publish, adapt, perform
Exceptions	Farmers' privilege, researcher's rights	Compulsory licensing, public interest provi-sions	Fair use provisions for education/re-search

5.3.6 Enforcement Mechanisms

PVP	Patent	Copyright
Civil suits for infringement, revocation proceedings	Civil and criminal remedies, opposition proceedings	Civil suits, statutory damages, injunctions
Tribunal of PPV&FR Authority holds jurisdiction	Indian Patent Office and Intellectual Property Appellate Board (IPAB)	Copyright Board handles disputes and registration processes

5.3.7 Key Use Cases in Biotechnology

Use Case	Best Protected By	Explanation
Genetically engineered seed variety	PVP + Patent (if process patented)	Variety under PVP; transformation method under Patent
CRISPR-based genome editing technique	Patent	Invention of method, guide RNA design, and applications
Bioinformatics software for sequence analysis	Copyright	Protects source code and user interface
Protocol manual for in vitro regeneration	Copyright	Protects expression of method, not the scientific idea itself
Traditional crop variety improved with selection	PVP (if DUS criteria met)	Protected as extant or farmer-developed variety under PPV&FR Act

5.3.8 Interplay and Complementarity of IPRs

In many practical cases, more than one IPR may be employed **complementarily** to protect a single biotechnology innovation. For example:

- A **transgenic cotton** variety may be protected by
  - **Patent** for the gene construct and transformation method,
  - **PVP** for the new plant variety, and
  - **Trademark** for the brand name under which seeds are sold.

This layering of rights enables **holistic protection** of innovations from multiple legal dimensions.

5.3.9 Challenges in Interpretation and Overlaps

- **Overlap between PVP and Patent:** Some countries allow patenting of plant varieties (e.g., USA), but India excludes plants per se from patentability under Section 3(j) of the Patents Act.
- **Ambiguities in Copyrighted Protocols:** The boundary between **original expression** and **scientific method** is not always clear.
- **Institutional Ownership vs. Individual Inventor:** Complexities in rights assignment within public-sector research institutions and private companies.

5.3.10 Conclusion

Understanding the **distinctions between PVP, Patent, and Copyright** is foundational for navigating the complex IPR ecosystem in biotechnology. Each type of intellectual property caters to different **forms of creativity and innovation**, from breeding new plant varieties and inventing novel biological tools to developing educational materials and bioinformatics software.

A strategic and informed use of the appropriate IPR tool—based on the **nature of innovation, regulatory frameworks**, and **commercial goals**—is essential to maximize **innovation incentives, legal protection**, and **societal benefit** in the age of biotechnological advancement.

6

Global Trade and IPR

6.1 World Trade Organization (WTO) and Trade-Related Aspects of Intellectual Property Rights (TRIPS)

6.1.1 Introduction

The globalization of trade and the rapid advancement in innovation—particularly in biotechnology—have necessitated the harmonization of **intellectual property rights (IPR)** across nations. The **World Trade Organization (WTO)**, established in 1995, is the apex multilateral body that regulates international trade among member states. Within its legal framework, the **Agreement on Trade-Related Aspects of Intellectual Property Rights (TRIPS)** plays a foundational role in setting **minimum standards for IPR protection** and aligning national laws with global trade norms.

In the realm of **biotechnology**, TRIPS has profound implications on **patentability of life forms, technology transfer, public health, biodiversity, and traditional knowledge**.

6.1.2 World Trade Organization (WTO): An Overview

- **Established:** January 1, 1995
- **Headquarters:** Geneva, Switzerland
- **Membership:** 164 members (as of 2024), including India
- **Primary Objectives:**
    - Facilitate free and fair global trade
    - Administer trade agreements
    - Resolve trade disputes
    - Monitor national trade policies
    - Promote cooperation with other international organizations (e.g., WIPO, WHO)

The WTO is built upon the foundation of the **General Agreement on Tariffs and Trade (GATT)** and encompasses key trade-related agreements, including **GATS (services)**, **TRIPS (IP rights)**, and **AoA (agriculture)**.

6.1.3 TRIPS Agreement: Origin and Evolution

- **Came into force:** January 1, 1995
- **Binding instrument** under the WTO legal system
- **Minimum Standards Agreement:** It sets out the **minimum level of protection** each member state must provide for IPRs.

Scope of TRIPS

- Copyrights and related rights
- Trademarks, including service marks
- Geographical indications
- Industrial designs
- Patents
- Layout-designs of integrated circuits
- Protection of undisclosed information (trade secrets)

Relevance to Biotechnology

TRIPS Article 27.3(b) addresses the controversial issue of **patenting life forms**, which has had far-reaching consequences for biotechnology research, agricultural innovation, and biodiversity conservation.

6.1.4 Key Provisions of TRIPS Relevant to Biotechnology

Provision	Description and Implication
Article 27 (Patentable Subject Matter)	Mandates patents for inventions in all fields of technology; allows exclusion of plants, animals, and essentially biological processes; however, **microorganisms and microbiological processes must be patentable**.
Article 27.3(b)	Nations must allow either **patents** or an **effective sui generis system** (e.g., PVP) for plant varieties. This is where India enacted the **PPV&FR Act** as a compliant mechanism.
Article 31 (Compulsory Licensing)	Allows governments to issue compulsory licenses in cases of national emergency, public health crisis, etc.—crucial for **biopharmaceuticals** and **agro-biotech inputs**.
Article 39 (Undisclosed Information)	Protection for **trade secrets** and **confidential data**, particularly in **clinical trials, biotechnology R&D**, and **regulatory dossiers**.

6.1.5 TRIPS Compliance and Indian Legislative Response

India, as a WTO member, is bound by TRIPS obligations. Consequently, the country has modified several laws to ensure **TRIPS compliance**, especially in the IPR domain.

TRIPS Requirement	Indian Legal Response
Patent protection for inventions	*The Patents Act, 1970* (amended in 1999, 2002, and 2005)
Plant variety protection	*Protection of Plant Varieties and Farmers' Rights (PPV&FR) Act, 2001*
Copyright standards	*The Copyright Act, 1957* (amended in 2012)
Protection of undisclosed data	*Drugs and Cosmetics Rules* and industry confidentiality norms

India chose not to allow **patents on plants and animals**, instead developing a sui generis system for PVP. This balances the TRIPS mandate with India's biodiversity, farmers' rights, and public interest.

6.1.6 Biotechnology, TRIPS, and Global Trade: Strategic Implications

1. **Innovation and Market Access**
   - TRIPS encourages investment in **biotechnological innovation** by ensuring protection in international markets.
   - Enables companies to seek **patent rights for bioproducts**, such as vaccines, enzymes, or GMOs, in WTO-compliant countries.
2. **Technology Transfer and R&D Collaboration**
   - Facilitates international **joint ventures**, **biotech alliances**, and **licensing agreements**, especially when strong IPR regimes are in place.
3. **Ethical and Sovereignty Concerns**
   - Debates around **biopiracy**, **patent monopolies**, and **corporate control over seeds and medicines** have emerged, leading to calls for balancing IP protection with **equity and access**.
4. **Public Health and Pandemic Response**
   - COVID-19 highlighted **TRIPS flexibilities** such as **compulsory licensing** and **waivers** for vaccines and diagnostic technologies.
5. **Agri-biotech and Farmer Rights**
   - Seed sovereignty and **indigenous knowledge systems** face new pressures under **TRIPS-aligned patent regimes**.
   - Countries like India emphasize a **TRIPS-plus framework** that incorporates **traditional knowledge protection**.

6.1.7 Challenges and Criticism of TRIPS

Issue	**Explanation**
Access vs. Innovation	Excessive patent protection may **restrict access** to essential technologies.
One-size-fits-all approach	Same standards applied to developed and developing countries, causing imbalance.
Biopiracy and Misappropriation	Traditional genetic resources and indigenous innovations not adequately protected.
Lack of clarity in Article 27.3(b)	Leaves space for varied national interpretations; periodic review remains unresolved.

6.1.8 Conclusion

The **TRIPS Agreement**, under the aegis of the **WTO**, plays a transformative role in aligning national IPR regimes with the needs of global trade. For biotechnology, TRIPS has opened opportunities for innovation, investment, and international collaboration, but also raised complex ethical, economic, and sovereignty issues.

India's legal strategy—through **balanced amendments**, a **sui generis PVP system**, and assertive use of **flexibilities**—provides a model for **equitable TRIPS compliance** while safeguarding national interests.

In future biotechnology policy-making, a nuanced understanding of TRIPS will remain essential for **IP management**, **innovation strategy**, and **global competitiveness**.

6.2 Cross-border Movement of Germplasm and Genetic Resources

6.2.1 Introduction

The **cross-border movement of germplasm and genetic resources** is a critical component of agricultural biotechnology, plant breeding, biodiversity conservation, and global food security. Germplasm, the living genetic material such as seeds, tissues, or reproductive cells used for breeding, research, and conservation, often needs to be shared internationally for scientific innovation and sustainable agriculture.

However, the international exchange of such genetic materials raises important legal, ethical, and geopolitical concerns, particularly in the context of **sovereign rights**, **intellectual property**, **bioprospecting**, and **benefit sharing**. These movements are now governed by a complex set of international treaties, national regulations, and institutional frameworks, aimed at ensuring **fair and equitable access and use**.

6.2.2 Importance of Germplasm Exchange in Biotechnology

- **Crop Improvement**: Access to diverse germplasm is essential for developing high-yielding, climate-resilient, pest-resistant crop varieties.
- **Conservation of Biodiversity**: Facilitates ex situ conservation and gene banking strategies.
- **Biotechnological Research**: Underpins molecular breeding, genomic studies, and transgenic development.
- **Global Food Security**: Ensures availability of genetic resources to address future agricultural challenges.

6.2.3 International Regulatory Frameworks Governing Germplasm Exchange

1. Convention on Biological Diversity (CBD), 1992

- **Principle of Sovereignty**: Recognizes national sovereignty over biological resources.
- **Prior Informed Consent (PIC)** and **Mutually Agreed Terms (MAT)**: Required before accessing genetic materials.
- **Access and Benefit-Sharing (ABS)**: Equitable sharing of benefits arising from the use of genetic resources.
- India enacted the **Biological Diversity Act, 2002** to comply with CBD principles.

2. Nagoya Protocol on Access and Benefit-Sharing, 2010

- Supplementary agreement to CBD.
- Provides a transparent legal framework for:
  - Access to genetic resources
  - Fair and equitable benefit sharing
  - Compliance by user countries
- Mandates disclosure of **source of origin** and adherence to **national ABS laws**.

3. International Treaty on Plant Genetic Resources for Food and Agriculture (ITPGRFA), 2001

- Recognizes **Plant Genetic Resources for Food and Agriculture (PGRFA)** as global commons.
- Developed the **Multilateral System (MLS)** for access to 64 key crops (Annex 1 crops).

- Facilitates
  - Easy access for research, breeding, and training
  - Standard Material Transfer Agreement (SMTA)
  - Benefit-sharing through financial or non-monetary means
- Operates under the FAO and complements CBD.

4. WTO-TRIPS Agreement (Article 27.3(b))

- Allows exclusion of plants and animals from patentability.
- Mandates effective protection of plant varieties through patents or a sui generis system.
- Indirectly influences national rules on access and control of genetic materials.

6.2.4 India's Policy and Legal Framework on Germplasm Exchange

Legal Instrument	Key Provisions and Relevance
Biological Diversity Act, 2002	Regulates access to biological resources and knowledge by foreigners. Mandates PIC and benefit sharing.
Protection of Plant Varieties and Farmers' Rights (PPV&FR) Act, 2001	Encourages conservation and recognizes farmers as conservers and breeders.
The Patents Act (Amended), 2005	Excludes plants and animals from patentability; allows patents for microbiological processes.
Plant Quarantine (Regulation of Import into India) Order, 2003	Ensures biosafety and phytosanitary control of imported germplasm.
National Biodiversity Authority (NBA)	Central authority responsible for granting approval for international germplasm exchange.

6.2.5 Challenges in Cross-border Movement of Germplasm

Challenge	Implication
Biopiracy	Unauthorized use of indigenous genetic resources or traditional knowledge without benefit-sharing.
Regulatory Complexity	Overlapping global and national laws often create procedural delays and bureaucratic hurdles.
Sovereignty Conflicts	Tension between countries of origin and user countries over access rights and benefit-sharing terms.
Limited Benefit Realization	Inadequate mechanisms to ensure fair returns to source countries or communities.
Data Disclosure and Traceability	Ensuring disclosure of source origin and monitoring usage of materials is challenging.

6.2.6 Case Examples of Germplasm Controversies and Resolutions

1. Neem Patent Case (India vs. W.R. Grace, Europe)

- A European patent on neem-based antifungal formulation was challenged by Indian scientists.
- The patent was revoked by the European Patent Office citing **lack of novelty** and prior indigenous knowledge.
- Highlighted the need for recognizing **traditional knowledge** and equitable benefit sharing.

2. Turmeric Patent Case

- US Patent granted on turmeric wound healing property.
- Indian scientists provided prior art evidence, leading to **revocation of the patent**.
- Became a global symbol of **biopiracy resistance**.

3. IRRI and Indian Rice Germplasm

- Concerns raised over **accession of basmati and traditional rice varieties** in international collections.
- India demanded **access logs, benefit-sharing**, and protection under **Geographical Indication (GI)**.

6.2.7 Best Practices for Germplasm Exchange

- Adherence to **Material Transfer Agreements (MTA)** or **SMTA** under ITPGRFA.
- Establishment of **Institutional Biosafety Committees (IBSCs)**.
- **Digital documentation** of genetic accessions through barcoding or genome sequencing.
- **Compliance audits** and traceability systems to monitor germplasm flow.
- Incorporation of **Traditional Knowledge Digital Libraries (TKDLs)** to prevent patent misappropriation.

6.2.8 Future Directions and Recommendations

- **Digital Sequence Information (DSI)** governance under ongoing international negotiations.
- Creation of **global ABS tracking systems** for enhanced transparency.
- Development of **open-access bioresource repositories** with fair licensing mechanisms.

- Empowering **indigenous communities** with IPR education and participatory governance.
- National strategy for **genetic sovereignty**, balanced with global innovation goals.

6.2.9 Conclusion

The **cross-border movement of germplasm and genetic resources** is indispensable for global scientific progress, agricultural sustainability, and climate adaptation. However, it must be governed with fairness, transparency, and respect for biodiversity and traditional rights. International conventions like **CBD, Nagoya Protocol, ITPGRFA**, and TRIPS, along with national legislations such as India's **BDA and PPV&FR Act**, together form a comprehensive legal framework to regulate such exchanges.

Balancing **innovation** with **sovereignty** and **equity** will be critical in shaping a sustainable and just global biotechnology landscape.

6.3 Ecological and Economic Implications of IPR on Commercialization

6.3.1 Introduction

The application of **Intellectual Property Rights (IPR)** in biotechnology has catalyzed innovation, investment, and commercialization of biological resources. However, the ecological and economic consequences of enforcing IPRs — particularly patents, plant breeder's rights, and trademarks — are increasingly debated. The **commercialization of biotech innovations** under the IPR regime affects not only industrial stakeholders but also **farmers, indigenous communities, biodiversity**, and **local economies**.

This section analyses the **dual-edged impacts** of IPRs: enabling commercialization and technology transfer, while also influencing ecological balance and socio-economic equity.

6.3.2 Economic Implications of IPR in Biotechnology Commercialization

1. Encouragement of R&D and Private Investment

- Strong IPR protection incentivizes biotech firms to invest in **long-term R&D**.
- Encourages development of novel products like **transgenic crops**, **biofertilizers**, and **biopharmaceuticals**.
- Facilitates **technology transfer and licensing** to small and medium enterprises (SMEs).

2. Market Monopoly and Pricing Issues

- Patent holders often acquire **monopoly rights**, enabling **premium pricing** of seeds, diagnostics, or therapeutics.
- Example: Bt cotton seeds and genetically modified (GM) soybean are priced significantly higher due to proprietary control.
- May **exclude resource-poor farmers** from accessing high-tech seeds or inputs.

3. Impact on Traditional Agriculture and Livelihoods

- Farmers lose the **freedom to save, exchange, or reuse seeds** when patented varieties are introduced.
- **Shift from informal to formal seed systems** increases dependency on commercial seed firms.
- Affects **sustainability of indigenous cropping systems** and traditional seed networks.

4. Revenue Generation and Benefit Sharing

- Royalties from patented products generate **national revenue**, especially where benefit-sharing mechanisms exist (e.g., PPV&FR Authority).
- Technology licensing creates **employment**, **exports**, and **spin-off industries**.

5. Biopiracy and Unequal Benefit Distribution

- IPR enforcement without proper access and benefit-sharing frameworks leads to **biopiracy** — misappropriation of genetic resources and traditional knowledge.
- Example: Patents on turmeric, neem, and basmati rice by foreign entities without benefit-sharing with Indian stakeholders.

6.3.3 Ecological Implications of IPR-Driven Commercialization

1. Genetic Uniformity and Biodiversity Loss

- Commercialization of **IP-protected, high-yielding monocultures** reduces **genetic diversity** in agroecosystems.
- Increases vulnerability to **pests, diseases, and climate stresses**.
- Contradicts global efforts toward **agro-biodiversity conservation**.

2. Biosafety and Environmental Risk

- Release of genetically modified organisms (GMOs) involves **ecological risks** such as:

  - Gene flow to wild relatives
  - Emergence of superweeds or resistant pests
  - Non-target organism effects
- Requires **rigorous environmental impact assessments (EIA)** and biosafety regulations.

3. Erosion of Indigenous Knowledge Systems

- Proprietary IPR systems often ignore or **undervalue traditional knowledge** and practices.
- Documentation and protection through **Traditional Knowledge Digital Libraries (TKDLs)** are still evolving.
- May lead to **cultural erosion and ecological mismanagement**.

4. Seed Sovereignty and Dependency

- Patented seed technologies can **disempower local farmers**, making them reliant on corporate seed supplies every season.
- Threatens the principle of **seed sovereignty**, particularly in developing countries.

5. Disruption of Ecological Balance

- Over-commercialization of a few genetically engineered crops may lead to **invasive species problems**, altered pollination patterns, and **soil microbiome disruption**.
- Example: Herbicide-tolerant crops affect weed flora and can indirectly affect soil fauna.

6.3.4 Balancing IPR and Ecological-Economic Equity

To ensure **sustainable commercialization** of biotechnology under IPR regimes, a **holistic and inclusive approach** is essential. Strategies may include:

Approach	Ecological and Economic Benefit
Sui generis systems (e.g., PPV&FR Act)	Encourage farmer participation and biodiversity conservation while allowing formal protection.
Compulsory Licensing Provisions	Enable access to essential biotech products in public interest, especially in health and agriculture.
Material Transfer Agreements (MTAs)	Ensure transparency and fairness in cross-border germplasm exchanges.
Community Participation in IPR Decisions	Recognize traditional knowledge holders and local innovators in benefit-sharing frameworks.

Environmental and Social Impact Assessments (ESIA)	Evaluate ecological consequences before product commercialization.
Open-Source Biotech Models	Facilitate innovation through shared platforms and data without proprietary constraints.

6.3.5 Case Insights: India and Global Scenarios

Case 1: Bt Cotton in India

- Introduced under patent and licensing agreements.
- Led to increased yields initially, but later issues included:
  - High seed cost
  - Bollworm resistance
  - Farmer indebtedness and suicides in some regions
- Highlighted the need for **local adaptation**, **public-sector breeding**, and **biosafety vigilance**.

Case 2: Golden Rice

- A genetically modified rice with pro-vitamin A trait, patented by several entities.
- Delayed adoption due to **biosafety**, **licensing complexity**, and **public resistance**.
- Underlines how **IPR complexity can delay life-saving innovations**.

Case 3: Indian Traditional Varieties and Patent Challenges

- Legal battles against turmeric and neem patents show how **community-driven IPR resistance** can preserve **ecological heritage** and **economic justice**.

6.3.6 Conclusion

The commercialization of biotechnology under the umbrella of IPRs must be pursued with **balanced foresight**. While IPRs are crucial for rewarding innovation and facilitating technology diffusion, **unregulated or narrowly designed IPR regimes** can create ecological vulnerabilities and economic exclusions.

India's pluralistic legal approach — combining **patents**, **plant variety protection**, **environmental safeguards**, and **community rights** — offers a template for **sustainable IPR frameworks** in biotechnology. Future policies must ensure that **economic progress through IPR** does not come at the cost of **ecological integrity** or **social equity**.

PART-II
Biosafety and Regulatory Frameworks

7

Fundamentals of Biosafety and Biohazards

7.1 Definitions and Classifications of Biohazards

7.1.1 Definition of Biohazards

Biohazards, or **biological hazards**, refer to biological substances that pose a threat to the health of living organisms, primarily humans, animals, and plants. These substances include a broad spectrum of infectious agents such as bacteria, viruses, fungi, protozoa, parasites, recombinant or synthetic nucleic acid molecules, genetically modified organisms (GMOs), and biological toxins. Biohazards can arise from research laboratories, agricultural practices, biotechnology industries, healthcare settings, or naturally occurring infectious diseases.

In biotechnology and molecular biology, biohazards are particularly significant due to the frequent use of genetically modified microbes, transgenic plants and animals, and gene-editing tools that may unintentionally cause ecological disturbances or health risks if not properly contained and managed.

7.1.2 Understanding Biosafety in the Context of Biohazards

Biosafety is the application of containment principles, facility design, practices, and procedures to prevent unintentional exposure to biohazards or their accidental release. It forms the foundational pillar in laboratories and institutions engaged in biotechnology, genetic engineering, and molecular biology.

The goal of biosafety is twofold

- **To protect laboratory personnel, the community, and the environment** from potential harm caused by biological agents.
- **To maintain scientific integrity and public trust** by ensuring safe practices in biotechnological innovations.

7.1.3 Classification of Biohazards by Risk Groups

The **World Health Organization (WHO)** and various national biosafety frameworks classify biohazardous agents into four **Risk Groups (RGs)** based on their pathogenicity, mode of transmission, availability of preventive and therapeutic measures, and the risk they pose to individuals and the community.

Risk Group	Definition	Examples
Risk Group 1 (RG1)	Agents not associated with disease in healthy adult humans.	*Escherichia coli* (non-pathogenic strain K-12), *Bacillus subtilis*
Risk Group 2 (RG2)	Agents associated with human disease which is rarely serious and for which preventive or therapeutic interventions are often available.	*Salmonella spp.*, *Mycobacterium tuberculosis* (some strains), *Adenoviruses*
Risk Group 3 (RG3)	Agents associated with serious or potentially lethal human diseases, for which preventive or therapeutic interventions may be available.	*Mycobacterium tuberculosis* (virulent strain), *HIV*, *SARS-CoV-2*
Risk Group 4 (RG4)	Agents likely to cause serious or lethal human diseases for which preventive or therapeutic interventions are not usually available.	*Ebola virus*, *Marburg virus*, *Lassa fever virus*

7.1.4 Types of Biohazards

Biohazards can be further classified based on their biological nature and origin. This classification is critical for determining the level of containment, the nature of safety precautions, and regulatory oversight.

1. **Infectious Agents**: Pathogens such as bacteria, viruses, fungi, and parasites capable of causing diseases in humans, animals, or plants. *Example*: *Bacillus anthracis*, *Hepatitis B Virus*, *Phytophthorainfestans*.
2. **Recombinant DNA and Genetically Modified Organisms (GMOs)**: Engineered organisms or DNA molecules that may possess altered characteristics with potential biosafety implications. *Example*: Herbicide- resistant transgenic crops, gene drive-modified insects.
3. **Biological Toxins**: Toxic substances produced by living organisms, including bacterial exotoxins, mycotoxins, and plant or animal toxins. *Example*: Botulinum toxin, Ricin.
4. **Allergens**: Biotechnological products that may induce hypersensitive reactions or allergies upon exposure. *Example*: Novel proteins expressed in GM foods.
5. **Prions**: Misfolded proteins that can induce abnormal folding of normal proteins, leading to degenerative diseases. *Example*: Prions causing

Creutzfeldt–Jakob disease (CJD) or Bovine Spongiform Encephalopathy (BSE).

7.1.5 Routes of Biohazard Exposure

Understanding the possible routes of exposure is vital for designing biosafety protocols:

- **Inhalation**: Aerosols or airborne transmission in laboratory settings.
- **Ingestion**: Accidental swallowing of infectious materials.
- **Percutaneous**: Needlestick injuries, cuts, or skin punctures.
- **Mucosal contact**: Splash into eyes, nose, or mouth.

7.1.6 Biohazards in the Biotechnology Sector

In modern agricultural and industrial biotechnology, biohazards may arise from:

- Use of live genetically engineered microbes in biofertilizers or biopesticides.
- Unintended gene flow from GM crops to wild relatives.
- Unregulated disposal of GM waste products.
- Unforeseen immune or allergic responses to transgenic food.

Therefore, recognizing and managing biohazards is indispensable in research, development, and commercialization of biotechnological innovations.

7.1.7 Conclusion

An accurate understanding of biohazards and their classifications forms the cornerstone of a safe biotechnological practice. With the increasing sophistication of molecular tools and synthetic biology, biosafety must evolve concurrently to ensure that the benefits of biotechnology are not overshadowed by unintended risks. The integration of sound biosafety principles, global regulatory frameworks, and ethical considerations is essential for sustainable and responsible biotech advancement.

7.2 Biosafety: Origin, Scope, and Importance in Biotechnology

7.2.1 Origin and Evolution of Biosafety

The concept of **biosafety** originated in response to the increasing use of infectious agents and genetically modified organisms (GMOs) in laboratory and industrial environments. The term gained prominence during the **1975 Asilomar Conference on Recombinant DNA**, where scientists and ethicists convened to discuss potential risks associated with recombinant DNA (rDNA) technology. The conference emphasized the need for containment measures, ethical guidelines, and precautionary approaches to manage biohazards arising from genetic engineering.

Key historical milestones include

- 1975: *Asilomar Conference* – formally initiated the discourse on biosafety guidelines for rDNA work.
- 1983: WHO issued its first *Laboratory Biosafety Manual.*
- 1990s: With the commercialization of GMOs, countries began establishing formal biosafety regulatory frameworks.
- 2000: The Cartagena Protocol on Biosafety came into force to regulate transboundary movement of living modified organisms (LMOs).

Thus, biosafety evolved as a multidisciplinary field encompassing **biological sciences, public health, policy, law, and ethics**.

7.2.2 Concept and Definition of Biosafety

Biosafety refers to the **principles, technologies, practices, and procedures** implemented to prevent unintentional exposure to pathogens and toxins, or their accidental release, particularly in the context of research involving GMOs, recombinant DNA, and hazardous biological agents.

It ensures

- The **safety of humans, animals, and plants** from laboratory-associated infections or environmental contamination.
- The **containment of harmful biological materials** in confined settings.
- The **ethical responsibility** of biotechnology in maintaining ecological and health integrity.

7.2.3 Scope of Biosafety in Biotechnology

Biosafety has become an **indispensable pillar** in modern biotechnology due to the expansive use of

- Transgenic crops and animals.
- Genetically modified microorganisms in industry.
- Synthetic biology and genome editing tools (e.g., CRISPR-Cas).
- Molecular diagnostics and vaccine development.

Its scope extends across

1. **Academic and Research Laboratories**: Ensuring safe handling of GMOs, viruses, and infectious cultures in educational institutions and R&D centers.

2. **Agricultural Biotechnology**: Implementing risk assessments and gene flow containment in field trials and release of GMOs.
3. **Industrial Biotechnology**: Regulating large-scale bioprocesses involving engineered microbes used for biofuels, enzymes, or pharmaceuticals.
4. **Medical Biotechnology**: Maintaining sterility and biosafety in vaccine production, recombinant drug formulation, and gene therapy.
5. **Environmental Biotechnology**: Preventing unintended ecological consequences of releasing GMOs for bioremediation or biosensing.
6. **Public Health Interface**: Strengthening preparedness against accidental or deliberate release of biohazards (biosurveillance, bioterrorism concerns).

7.2.4 Importance of Biosafety in Biotechnology

The significance of biosafety in biotechnology lies in its ability to address both **scientific risks** and **societal concerns**. Key importance includes:

1. **Risk Mitigation**: Prevents occupational exposure, environmental contamination, and unintended horizontal gene transfer.
2. **Public Trust and Transparency**: Ensures community acceptance and regulatory compliance, particularly in contentious areas like GM foods.
3. **Policy and Regulation Support**: Acts as the scientific foundation for biosafety policies at national (e.g., RCGM, GEAC) and international (e.g., Cartagena Protocol) levels.
4. **Ethical Research Practices**: Encourages responsible innovation by integrating ethical, ecological, and health perspectives.
5. **International Trade and Compliance**: Facilitates the export/import of biotech products under agreed safety standards, avoiding trade barriers.
6. **Capacity Building and Training**: Enhances institutional readiness through personnel training, biosafety officer certification, and infrastructure development (e.g., BSL labs).

7.2.5 Biosafety vs. Biosecurity

While **biosafety** focuses on **preventing unintentional exposure** to biohazards, **biosecurity** deals with **preventing intentional misuse** of biological agents and technologies. In the biotechnology context, both are intertwined to create a robust safety net.

Aspect	Biosafety	Biosecurity
Objective	Prevent accidental release or exposure	Prevent misuse, theft, or sabotage
Focus	Human and environmental protection	National and global security
Relevance	R&D, agriculture, healthcare	Bioterrorism, dual-use research

7.2.6 Institutional and Regulatory Perspectives

In India, the biosafety framework is governed by the **Rules, 1989** of the Environment Protection Act (1986), and key regulatory bodies include:

- **Institutional Biosafety Committees (IBSCs)** – oversee project-level biosafety.
- **Review Committee on Genetic Manipulation (RCGM)** – monitors research and trial stages.
- **Genetic Engineering Appraisal Committee (GEAC)** – responsible for large-scale release and commercialization.

Global biosafety frameworks are coordinated through

- **WHO** (World Health Organization)
- **OECD** (Organisation for Economic Co-operation and Development)
- **UNEP-GEF** (Global Environment Facility)
- **Cartagena Protocol on Biosafety**

7.2.7 Conclusion

Biosafety represents a cornerstone of ethical, responsible, and sustainable biotechnological advancement. In an era where gene editing, synthetic biology, and molecular innovations are rapidly advancing, robust biosafety systems are essential not only to ensure **scientific integrity and environmental stewardship** but also to build **public confidence and global cooperation**. As biotechnology reshapes agriculture, medicine, and industry, biosafety will remain pivotal in securing a safe and beneficial biotechnological future.

7.3 Key Principles of Biosafety for Laboratory and Environmental Protection

7.3.1 Introduction

Biosafety principles are the **foundation of safe practices** in biotechnology laboratories and field environments. They ensure that potentially hazardous biological agents and genetically modified organisms (GMOs) are handled, contained, and disposed of in a manner that **prevents harm to personnel, the public, and the environment**. The objective is dual: protection of **human health** and preservation of **ecological integrity**.

The key biosafety principles are guided by **international protocols**, **national legislation**, and **scientific best practices**, forming the operational basis for all biotechnology-related activities across research, diagnostics, agriculture, medicine, and industrial applications.

7.3.2 Core Principles of Laboratory Biosafety

1. **Risk Assessment**
   - **Definition**: Systematic evaluation of the probability and severity of adverse effects resulting from exposure to biohazardous agents.
   - **Implementation**: Involves identifying hazards, characterizing potential exposure, assessing consequences, and implementing control measures.
   - **Application**: Essential before starting work with GMOs, pathogens, or recombinant DNA.
2. **Containment**
   - **Primary Containment**: Achieved through safety equipment such as **biosafety cabinets (BSCs)**, sealed centrifuge cups, and proper use of personal protective equipment (PPE).
   - **Secondary Containment**: Structural features of laboratory design such as **airflow systems**, **autoclaves**, and **airlocks** that prevent escape of hazardous materials.
   - **Goal**: Prevent exposure within the lab and release into the external environment.
3. **Good Laboratory Practices (GLP)**
   - Standard procedures for handling, storage, labeling, waste disposal, decontamination, and sterilization.
   - Promotes **safety, reproducibility, and regulatory compliance**.
   - Includes maintenance of **biosafety manuals**, **training**, and **record-keeping**.
4. **Personnel Training and Competency**
   - All staff must receive formal training on **biosafety protocols**, hazard awareness, **emergency response**, and the use of containment devices.
   - Ongoing refresher programs are essential to maintain competency and awareness.
5. **Medical Surveillance and Health Monitoring**
   - Regular health checks, immunizations (e.g., Hepatitis B, Rabies), and incident documentation ensure **occupational safety**.

- Pre- and post-exposure monitoring are critical when working with high-risk agents.

6. **Incident Reporting and Response**
   - Immediate documentation and reporting of **accidents, spills, or exposures** is mandatory.
   - Incident response protocols must include **first aid**, **containment**, **disinfection**, and **regulatory reporting**.
7. **Waste Management and Decontamination**
   - Biological waste must be sterilized (e.g., autoclaved) before disposal.
   - Chemical disinfectants (e.g., sodium hypochlorite, ethanol) must be properly selected based on the biohazard class.
   - Decontamination protocols reduce the risk of accidental environmental exposure.
8. **Use of Biosafety Levels (BSL)**
   - Laboratories are classified into four **biosafety levels (BSL-1 to BSL-4)** depending on the pathogenicity of agents used:
     - **BSL-1**: Non-pathogenic agents (e.g., *E. coli* K12)
     - **BSL-2**: Moderate-risk agents (e.g., *Salmonella*, HIV)
     - **BSL-3**: High-risk airborne pathogens (e.g., *Mycobacterium tuberculosis*)
     - **BSL-4**: Life-threatening agents (e.g., Ebola virus)
   - Each level prescribes specific **containment, facility, and procedural standards**.

7.3.3 Biosafety Principles for Environmental Protection

1. **Environmental Risk Assessment (ERA)**
   - Evaluates the **potential ecological impact** of releasing GMOs or bio-products.
   - Factors considered:
     - Gene flow to wild species.
     - Effects on non-target organisms.
     - Soil and water ecosystem changes.
     - Resistance development in pests or pathogens.

2. **Gene Containment Strategies**
   - **Biological Containment**: Use of male sterility, seed terminator technology (e.g., GURTs), and chloroplast transformation to restrict gene flow.
   - **Physical Containment**: Isolation distances, buffer zones, and temporal separation (e.g., staggered sowing dates).
3. **Monitoring and Post-release Surveillance**
   - Periodic evaluation of environmental changes post-GMO release.
   - Helps identify unintended consequences such as biodiversity loss, allergenicity, or invasiveness.
4. **Emergency Preparedness and Response**
   - Protocols to handle unintended environmental release or contamination.
   - Includes **recall mechanisms**, **public communication**, and **mitigation strategies**.
5. **Regulatory Compliance and Reporting**
   - All GMO trials and releases must be pre-approved by regulatory bodies (e.g., RCGM, GEAC in India).
   - Regular audits and environmental reports are mandatory under **biosafety rules (1989)**.
6. **Public Engagement and Transparency**
   - Inclusion of **stakeholders, farmers, and communities** in biosafety decisions enhances acceptance.
   - Open access to biosafety data builds trust and ensures democratic governance of biotechnology.

7.3.4 Institutional Implementation of Biosafety Principles

In India, these biosafety principles are implemented through:

- **Institutional Biosafety Committees (IBSC)** – lab-level monitoring and approval.
- **Review Committee on Genetic Manipulation (RCGM)** – risk assessment and field trial oversight.
- **Genetic Engineering Appraisal Committee (GEAC)** – clearance for commercial release and environmental safety.

Globally, guidelines are harmonized by

- **World Health Organization (WHO)** Biosafety Manuals.
- **OECD Guidelines** for Recombinant DNA safety.
- **Cartagena Protocol on Biosafety** for transboundary GMO movement.

7.3.5 Conclusion

The principles of biosafety are integral to the **safe, ethical, and sustainable advancement of biotechnology**. Through rigorous containment, risk assessment, regulatory compliance, and transparent practices, biosafety ensures that the benefits of biotechnology are realized **without compromising human or ecological health**. As biotechnology becomes increasingly pervasive, a firm adherence to biosafety principles is not just a scientific imperative but a social and environmental necessity.

8

Biosafety Risk Assessment and Containment Strategies

8.1 Steps in Biosafety Risk Assessment

8.1.1 Introduction

Biosafety Risk Assessment is a **systematic, science-based process** used to identify, evaluate, and mitigate risks associated with handling biological agents, genetically modified organisms (GMOs), and biotechnological products. It is central to ensuring **human safety**, **environmental integrity**, and **regulatory compliance** in both laboratory and field applications.

A well-conducted biosafety risk assessment enables institutions to **establish appropriate biosafety levels**, implement **containment measures**, and ensure **sustainable and responsible use** of biotechnological advancements.

8.1.2 Key Objectives of Biosafety Risk Assessment

- Prevent exposure of personnel to biohazards.
- Avoid environmental contamination and unintended gene flow.
- Determine containment requirements for biological agents.
- Guide institutional and national biosafety decision-making.
- Support regulatory frameworks and compliance obligations.

8.1.3 Stepwise Framework for Biosafety Risk Assessment

The risk assessment process involves a **stepwise, iterative approach**, applicable across laboratory research, field trials, industrial bioprocessing, and agricultural biotechnology. The standard steps are as follows:

Step 1: Hazard Identification

Definition: Identification of all biological materials and processes that pose potential risks.

- Identify the organism or biological product (e.g., recombinant bacteria, transgenic plants, viruses).
- Evaluate whether it is **pathogenic, toxic, allergenic**, or **invasive**.

- Consider the biological nature: viability, infectivity, replication, or horizontal gene transfer.
- Classify agents according to national/international risk group (RG1–RG4).

Example: Use of *Agrobacterium tumefaciens* with a modified T-DNA vector in plant transformation warrants hazard identification for both the vector and the host organism.

Step 2: Hazard Characterization

Definition: Assessment of the **nature and magnitude** of potential adverse effects.

- Consider severity of harm: illness, mortality, environmental damage.
- Assess host range, transmission modes, and stability of the agent.
- Evaluate gene expression products (e.g., toxins, resistance genes).
- Include synergistic effects in multi-gene or multi-organism systems.

Example: Expression of a Bt toxin gene in cotton must consider non-target effects on beneficial insects like pollinators and predators.

Step 3: Exposure Assessment

Definition: Evaluation of how and to what extent people or the environment may come into contact with the hazard.

- Examine routes of exposure: inhalation, ingestion, dermal contact, accidental release.
- Analyze exposure duration, frequency, and population affected (lab personnel, farmers, ecosystem).
- Account for containment breach scenarios and escape probabilities.

Example: In field trials of genetically engineered rice, exposure to nearby wild rice relatives must be assessed for pollen-mediated gene flow.

Step 4: Risk Characterization

Definition: Integration of hazard and exposure information to estimate **overall biosafety risk**.

- Combine likelihood of harm with the consequences of exposure.
- Classify the risk level: negligible, low, moderate, or high.
- Assign the corresponding **biosafety level (BSL)** or containment protocol.
- Include uncertainty analysis to address gaps in data.

Example: A recombinant *E. coli* strain expressing an antibiotic resistance marker may be classified as moderate risk in a BSL-2 facility due to possible horizontal gene transfer.

Step 5: Risk Management Decisions

Definition: Formulation and application of **risk mitigation strategies** based on risk characterization.

- Implement administrative controls (e.g., SOPs, training).
- Apply engineering controls (e.g., biosafety cabinets, physical containment).
- Introduce biological barriers (e.g., sterility, kill switches, gene confinement).
- Enforce regulatory compliance (e.g., GEAC clearance, institutional biosafety committee approvals).

Example: In a confined field trial of GM mustard, regulators may mandate isolation distances and buffer zones to prevent cross-pollination.

Step 6: Risk Communication and Review

Definition: Transparent communication of findings to all stakeholders and updating risk assessments as new data emerge.

- Document all assumptions, decisions, and residual risks.
- Communicate with researchers, regulatory bodies, and the public.
- Reassess periodically or upon any change in protocol, environment, or organism behavior.

Example: A new report indicating unexpected insect resistance to Bt cotton should trigger a revision of the original risk assessment.

8.1.4 Factors Influencing Risk Assessment

- **Nature of the host and donor organisms**.
- **Genetic elements used** (promoters, selectable markers).
- **Gene expression products** (toxins, allergens, secondary metabolites).
- **Scale and duration** of the activity (laboratory, pilot, or commercial scale).
- **Environmental factors** (climate, biodiversity, presence of wild relatives).
- **Community sensitivity and ethical concerns**.

8.1.5 Role of Regulatory and Institutional Bodies

In India, risk assessments are mandated under the **Rules, 1989 of the Environment (Protection) Act, 1986**, and coordinated through:

- **Institutional Biosafety Committees (IBSCs)** – Risk assessment at the lab level.
- **Review Committee on Genetic Manipulation (RCGM)** – Scientific evaluation of GMO research.
- **Genetic Engineering Appraisal Committee (GEAC)** – National approval for environmental release.
- **State Biosafety Coordination Committees (SBCCs)** and **District Level Committees (DLCs)** – Monitoring field-level compliance.

8.1.6 Conclusion

Biosafety risk assessment is an **indispensable pillar** of responsible biotechnology. It ensures the **scientific rigor, environmental prudence, and ethical integrity** of biological research and innovation. By identifying and managing potential risks, this process protects human health, sustains biodiversity, and upholds public confidence in biotechnological progress. As biotechnology evolves, continuous improvement and dynamic reassessment of risks remain vital.

8.2 Biosafety Levels (BSL-1 to BSL-4) and Laboratory Containment

8.2.1 Introduction to Biosafety Levels (BSLs)

Biosafety Levels (BSLs) are a set of protective measures and laboratory practices designed to mitigate risks associated with handling biological agents. These levels are classified from **BSL-1 (lowest risk)** to **BSL-4 (highest risk)** based on:

- The **infectivity**, **pathogenicity**, and **transmissibility** of the microorganism.
- The **nature of the work** being performed (diagnostic, research, production).
- The **route of exposure** (inhalation, ingestion, percutaneous, mucosal).
- **Preventive and therapeutic interventions** available (vaccines, antibiotics).

Each level prescribes **specific laboratory infrastructure**, **safety equipment**, and **operational protocols** to contain hazards, thereby ensuring safety of personnel, community, and the environment.

8.2.2 Biosafety Level 1 (BSL-1): Basic Containment

Applicable to

Work involving **non-pathogenic**, well-characterized agents not known to cause disease in healthy adults.

Examples of Agents

- *Escherichia coli* K-12
- *Bacillus subtilis*
- *Saccharomyces cerevisiae*

Laboratory Features

- No special containment equipment required.
- Open bench work is permitted.
- Hand washing stations mandatory.
- Access to the lab is limited during work.

Practices

- Standard microbiological practices.
- Use of PPE like lab coats and gloves.
- Decontamination of work surfaces.

Containment Principle

Good laboratory practice (GLP) is the primary barrier.

8.2.3 Biosafety Level 2 (BSL-2): Moderate Risk Containment

Applicable to

Agents that pose **moderate hazards** to personnel and environment. These can cause human disease, but preventive or therapeutic measures are available.

Examples of Agents

- *Salmonella spp.*
- *Hepatitis B virus (HBV)*
- *Mycobacterium tuberculosis* (non-aerosol generating procedures)
- *Herpes simplex virus*

Laboratory Features

- Restricted access during operations.
- Biological Safety Cabinets (BSC) Class I or II for aerosol-generating tasks.

- Autoclaves for waste decontamination.
- Handwashing sinks and eyewash stations.

Practices

- All BSL-1 practices, plus:
- Limited lab access and signage.
- Proper sharps disposal.
- PPE includes face protection and gloves.

Containment Principle

Primary barriers (BSCs and PPE) combined with **procedural controls**.

8.2.4 Biosafety Level 3 (BSL-3): High Risk Containment

Applicable to

Agents that can cause **serious or potentially lethal infections** through inhalation. These are either indigenous or exotic and pose high risk to lab personnel and possibly the community.

Examples of Agents

- *Mycobacterium tuberculosis* (aerosol-generating procedures)
- *Yersinia pestis* (plague)
- SARS-CoV-2 (under certain conditions)
- *Francisellatularensis*

Laboratory Features

- Controlled and restricted access.
- Directional airflow with negative air pressure.
- Exhaust air not recirculated.
- BSC Class II or III mandatory.
- Sealed windows and walls.

Practices

- Medical surveillance and immunizations.
- Use of respirators where required.
- Strict lab entry protocols and record-keeping.

Containment Principle

Primary and secondary barriers, combined with **engineering controls**, ensure minimal exposure.

8.2.5 Biosafety Level 4 (BSL-4): Maximum Containment

Applicable to

Work with **dangerous and exotic agents** that pose high risk of life-threatening disease, for which there are no vaccines or treatments. Often involve **aerosol transmission** and high mortality rates.

Examples of Agents

- Ebola virus
- Marburg virus
- Lassa fever virus
- Crimean-Congo hemorrhagic fever virus

Laboratory Features

- Isolated, self-contained facility with dedicated air and water systems.
- BSL-4 labs may be:
    - **Cabinet laboratories**: All work within Class III BSCs.
    - **Suit laboratories**: Personnel wear full-body positive-pressure suits with air supply.
- Chemical shower for personnel before exiting.
- Double-door autoclaves and HEPA-filtered exhaust.

Practices

- Strict access control with biometric identification.
- Continuous monitoring and alarms for containment breach.
- No contact with external environment without full sterilization.

Containment Principle

Highest level of physical and biological containment, with **full isolation** from the external environment.

8.2.6 Laboratory Containment Strategies

Containment in biosafety refers to **methods used to manage infectious agents** to prevent exposure and dissemination. It includes:

A. Primary Containment

- Protects personnel and immediate lab environment.
- Achieved via PPE, BSCs, closed centrifuge rotors.

B. Secondary Containment

- Protects the community and environment outside the lab.
- Achieved via facility design (airflow, autoclaves, sealed rooms).

C. Engineering Controls

- HVAC systems with negative pressure.
- HEPA filtration and decontamination units.
- Alarmed doors, sealed penetrations, backup power.

D. Administrative Controls

- Biosafety training, SOPs, incident reporting, medical surveillance.

8.2.7 BSL Assignments for Biotechnology Applications

Application Area	Typical Biosafety Level
Recombinant DNA cloning using non-pathogenic hosts	BSL-1
GMOs expressing antibiotic resistance	BSL-2
Human pathogens in vaccine production	BSL-3
Work on hemorrhagic fever viruses	BSL-4

8.2.8 Regulatory Oversight in India

In India, biosafety level classification and enforcement are governed by:

- **Institutional Biosafety Committees (IBSC)** – assign lab containment levels.
- **Review Committee on Genetic Manipulation (RCGM)** – monitors BSL-2 and BSL-3 activities.
- **Genetic Engineering Appraisal Committee (GEAC)** – final approval for large-scale or BSL-4 related activities.
- **Guidelines issued by DBT (2017), MoEF&CC, and WHO** are followed.

8.2.9 Conclusion

Biosafety Levels (BSL-1 to BSL-4) are central to ensuring **safe biotechnological research and development**. Each level requires appropriate **infrastructure, training, and operational discipline**, aligned with the risk posed by biological materials. Establishing the correct biosafety level ensures **protection of researchers, the environment, and public health**, thereby reinforcing the ethical and sustainable application of biotechnology.

8.3 Handling, Storage and Disposal of Biohazardous Materials

8.3.1 Introduction

Biohazardous materials refer to **biological substances that pose potential threats to human health, animal health, or the environment**, including pathogenic microorganisms, recombinant DNA (rDNA) constructs, genetically modified organisms (GMOs), infectious agents, contaminated sharps, and human or animal tissue specimens. In biotechnology laboratories, proper **handling, storage, and disposal** of such materials is a critical biosafety mandate to prevent exposure, infection, contamination, and environmental release.

This section outlines the standard principles, regulatory protocols, and practical measures for safe management of biohazardous materials.

8.3.2 Handling of Biohazardous Materials

Proper handling is the first line of defense in minimizing biosafety risks. It involves a combination of **personal, procedural, and infrastructural precautions**:

A. Standard Operating Procedures (SOPs)

- All laboratory procedures involving biohazardous materials must follow written SOPs.
- SOPs must include information on risk classification, PPE requirements, emergency response, and decontamination steps.

B. Personal Protective Equipment (PPE)

- Minimum requirements include gloves, lab coats/gowns, and eye/face protection.
- BSL-3 and BSL-4 labs may require respirators or full-body suits.
- PPE must be donned and doffed under strict protocol to prevent cross-contamination.

C. Work Practice Controls

- Minimize aerosol generation (e.g., avoid vigorous pipetting, vortexing outside biosafety cabinets).
- Use of **Biological Safety Cabinets (BSCs)** Class II or III for all open manipulations at BSL-2 and above.
- Label all containers with **biohazard symbols**, indicating material class and handling instructions.

D. Decontamination Protocols

- Surfaces must be disinfected using appropriate agents (e.g., 10% sodium hypochlorite, 70% ethanol, phenolics).
- Equipment (centrifuges, incubators) must be routinely decontaminated.
- Autoclaving is the standard method for sterilizing biohazardous tools and waste.

8.3.3 Storage of Biohazardous Materials

Secure and environmentally appropriate storage is necessary to maintain containment integrity and prevent accidental exposure or release.

A. Labeling and Documentation

- All storage containers must be clearly labeled with:
  - Name and strain of the organism
  - Biosafety level
  - Date of storage and responsible personnel
- Inventory logs must be meticulously maintained and routinely audited.

B. Containment Infrastructure

- Refrigerators and freezers designated for biohazards should:
  - Be lockable and separated from general-use units.
  - Include secondary containment trays to prevent spillage.
  - Have temperature monitoring and alarm systems for sensitive materials.

C. Segregation and Zoning

- Separate zones for different biosafety levels.
- Segregation of viable materials from non-viable or non-infectious samples.

D. Cryopreservation and Long-Term Storage

- Viable pathogens and cell lines may be stored in cryogenic conditions.
- Cryovials must be securely capped and stored in liquid nitrogen tanks with explosion-proof shielding.
- Periodic viability and contamination checks are mandatory.

8.3.4 Disposal of Biohazardous Materials

Disposal must comply with **national biosafety and biomedical waste disposal rules** to prevent hazards to lab workers, waste handlers, and the environment.

A. Classification of Waste

- **Sharps** (needles, blades, glass): Puncture-proof containers with biohazard label.
- **Contaminated solids** (gloves, petri dishes, tubes): Autoclaved and incinerated or sent to authorized biomedical waste handlers.
- **Liquid cultures and supernatants**: Chemically disinfected (e.g., sodium hypochlorite, glutaraldehyde) and then disposed of through drains as per protocol.
- **Animal carcasses and tissue**: Deep burial or incineration under supervision.

B. Decontamination Methods

- **Autoclaving**: Most effective method (121°C, 15 psi, 15–60 minutes depending on material).
- **Chemical disinfection**: For materials incompatible with autoclave (e.g., plastics, liquid waste).
- **Incineration**: For high-risk waste, particularly BSL-3/4 waste and carcasses.

C. Waste Segregation and Color Coding (as per Biomedical Waste Rules, India)

Type of Waste	Color-Coded Bin	Treatment/Disposal
Contaminated plastics/gloves	Red	Autoclaving & shredding
Sharps (needles, glass)	White (translucent)	Autoclaving & encapsulation
Infectious biological waste	Yellow	Incineration or deep burial
Broken non-contaminated glass	Blue	Disinfection & recycling

D. Documentation and Record-Keeping

- Logs of biohazard disposal, autoclave cycles, and incineration must be maintained for inspection by biosafety authorities and audit teams.

E. Authorized Disposal Agencies

- Institutions must collaborate with **Common Biomedical Waste Treatment and Disposal Facilities (CBWTF)** as per Central Pollution Control Board (CPCB) and Ministry of Environment, Forest and Climate Change (MoEF&CC) guidelines.

8.3.5 Emergency Management for Accidents and Spills

- Spill kits must be available in all biosafety zones.
- Immediate containment, neutralization with disinfectants, and proper reporting are mandatory.

- Personnel must undergo **biosafety incident response training**.
- First aid, exposure reporting, and medical evaluation procedures should be established and tested periodically.

8.3.6 Regulatory Guidelines in India and Internationally

- **India**
  - **DBT Biosafety Guidelines (2017)**.
  - **Biomedical Waste Management Rules (2016, amended 2018)** under MoEF&CC.
  - Institutional Biosafety Committees (IBSCs) monitor compliance.
- **Internationally**
  - **WHO Laboratory Biosafety Manual (4th Edition)**.
  - **NIH Guidelines** for Research Involving Recombinant or Synthetic Nucleic Acid Molecules (USA).
  - **OSHA and CDC Biosafety Standards**.

8.3.7 Conclusion

The safe **handling, storage, and disposal** of biohazardous materials is a cornerstone of laboratory biosafety and environmental stewardship in biotechnology. It involves a **multi-layered system of technical practices, infrastructure controls, waste management procedures, and legal compliance**. Institutions must invest in biosafety training, infrastructure, and audit mechanisms to uphold bioethical standards and ensure sustainable scientific progress.

8.4 Case Studies of Biosafety Lapses

8.4.1 Introduction

Biosafety lapses—whether due to human error, inadequate infrastructure, or policy non-compliance—have historically caused **laboratory-acquired infections (LAIs)**, **unintentional environmental releases**, and **serious public health risks**. Case studies of such incidents offer critical lessons in improving biosafety governance and laboratory practices. This section presents a selection of internationally reported biosafety failures and their implications, with a focus on their relevance to biotechnology and genetic research facilities.

8.4.2 Case Study 1: SARS Virus Escapes from Laboratories (Singapore and China, 2003–2004)

Context

Following the 2002–2003 global outbreak of Severe Acute Respiratory Syndrome (SARS-CoV), researchers stored and studied live viral cultures under BSL-3 conditions. However, subsequent **accidental laboratory exposures** led to new infections.

Incidents

- In **Singapore (2003)**, a graduate student working in a BSL-3 laboratory became infected. Investigation revealed **cross-contamination** between SARS-CoV and West Nile virus stocks due to improper cleaning of shared equipment.
- In **Beijing, China (2004)**, two lab workers at the **National Institute of Virology** became infected with SARS-CoV. One case led to secondary infections, resulting in **one death**.

Failures Identified

- Poor training on biosafety procedures.
- Inadequate equipment decontamination.
- Weak institutional oversight and auditing.

Key Takeaways

- Even advanced facilities are vulnerable to containment failure without strict **biosafety protocol adherence**.
- Biosafety audits and training must be continuous and rigorous.

8.4.3 Case Study 2: Accidental Distribution of H2N2 Virus by a U.S. Lab (2005)

Context:

In 2005, a U.S.-based laboratory inadvertently included **live H2N2 influenza virus**, the same strain responsible for the 1957–58 pandemic, in proficiency test kits sent to over **3,700 laboratories** worldwide.

Failures Identified

- Lack of biosafety review of test materials before distribution.
- Inadequate risk assessment regarding revival of eradicated strains.

Outcomes

- Global recall was issued by the **World Health Organization (WHO)**.
- Emergency destruction protocols were implemented by recipient labs.

Key Takeaways

- All pathogens, especially historical or eradicated ones, must undergo strict risk categorization.
- Laboratory proficiency testing must be governed by **external biosafety validation**.

8.4.4 Case Study 3: Bacillus anthracis Exposure at CDC (Atlanta, USA, 2014)

Context

At the **Centers for Disease Control and Prevention (CDC)**, up to **84 staff members were potentially exposed** to live *Bacillus anthracis* spores due to failure in confirming complete inactivation.

Failures Identified

- Lapses in biosafety level downgrading (BSL-3 to BSL-2) without confirmation.
- Inadequate documentation and procedural oversight.

Remedial Measures

- Temporary closure of labs.
- Personnel placed under antibiotic prophylaxis and health monitoring.
- Comprehensive biosafety reforms were instituted at CDC.

Key Takeaways

- **Verification of pathogen inactivation** is mandatory before transfer to lower containment zones.
- Biosafety practices must be periodically reviewed for procedural drift.

8.4.5 Case Study 4: Genetically Modified Wheat Field Contamination in Oregon, USA (2013)

Context

Unauthorized **genetically modified (GM) wheat**, developed by Monsanto, was discovered growing in a field in Oregon, though no GM wheat had been approved for commercial cultivation.

Implications

- Raised global concerns over unintentional GMO release.
- Japan and South Korea temporarily halted wheat imports from the U.S.

Failures Identified

- Improper containment during field trials.
- Weak post-trial site monitoring.

Key Takeaways

- **Field trials of GMOs must adhere to strict spatial and reproductive containment**.

- Post-trial surveillance is critical for preventing environmental persistence or spread.

8.4.6 Case Study 5: Foot-and-Mouth Disease Virus Escape from Pirbright Laboratory, UK (2007)

Context

The **Pirbright research facility**, which housed both vaccine and live virus stocks of Foot-and-Mouth Disease Virus (FMDV), was the origin of a localized FMD outbreak.

Findings

- The virus spread via a **damaged drainage pipe**, contaminating nearby farmland.
- The leak occurred from shared infrastructure between public research and private companies.

Consequences

- The UK government launched a full-scale inquiry.
- Biosafety infrastructure investments and laboratory segregation were recommended.

Key Takeaways

- **Facility maintenance and infrastructure auditing** are just as vital as procedural biosafety.
- **Public-private collaboration in biotech must enforce shared safety accountability**.

8.4.7 Lessons Learned from Biosafety Lapses

Across all case studies, several **recurring biosafety vulnerabilities** emerge:

Issue	Impact
Inadequate training	Human exposure, protocol breaches
Infrastructure failure	Environmental release, cross-contamination
Poor containment or inactivation	Transfer of live pathogens, infection risks
Weak regulatory oversight	Repeat offenses, global trust deficit
Lack of emergency preparedness	Delayed response, escalated consequences

8.4.8 Recommendations for Future Biosafety Governance

- **Regular training and certification** of personnel.
- **Strict inventory control and documentation** of hazardous biological materials.

- **Third-party biosafety audits** for high-containment facilities.
- **Enhanced global collaboration** through WHO, OECD, and FAO biosafety frameworks.
- Integration of **digital biosafety compliance tracking systems** (e.g., e-Biosafety Logs).

8.4.9 Conclusion

Biosafety lapses serve as cautionary tales underscoring that **technical knowledge alone is not sufficient**. Sustainable biosafety culture must be built upon **continuous risk assessment, robust infrastructure, accountable management, and ethical research practices**. These case studies form essential educational resources for shaping the future of responsible biotechnology.

9

Biosafety Policies and Legal Instruments

9.1 The Cartagena Protocol on Biosafety: Objectives and Implementation

9.1.1 Introduction

The **Cartagena Protocol on Biosafety** (CPB) is a pivotal international legal instrument under the **Convention on Biological Diversity (CBD)**, dedicated to regulating the **transboundary movement, handling, and use of living modified organisms (LMOs)** derived through modern biotechnology that may adversely affect biological diversity, human health, or the environment. Adopted in January 2000 in Cartagena, Colombia, and entering into force on **September 11, 2003**, the Protocol represents a **global consensus on biosafety governance** amidst the rapid expansion of genetic engineering technologies.

9.1.2 Objectives of the Cartagena Protocol

The principal objective of the Cartagena Protocol is:

"To contribute to ensuring an adequate level of protection in the field of the safe transfer, handling and use of living modified organisms resulting from modern biotechnology that may have adverse effects on the conservation and sustainable use of biological diversity, taking also into account risks to human health, and specifically focusing on transboundary movements." (Article 1, CPB)

The core objectives can be summarised as

- **Ensuring biosafety** in the context of international trade of LMOs.
- **Protecting biodiversity** from potential risks posed by LMOs.
- **Safeguarding human health** against unintended exposure to transgenic organisms.
- **Promoting informed decision-making** based on scientific risk assessment.
- **Upholding the sovereign rights of nations** to regulate access to their biological resources.

9.1.3 Scope and Applicability

The Protocol applies to **all LMOs** that may be transferred across borders, especially those intended for:

- **Introduction into the environment** (e.g., seeds, GM fish, transgenic insects).
- **Contained use** (e.g., laboratory research, bioreactors).
- **Food, feed, or processing (FFP)**—though these are regulated differently from LMOs for release into the environment.

It excludes

- **Non-living GM products** (e.g., processed food).
- LMOs used as pharmaceuticals for humans, which are regulated under the **World Health Organization (WHO)** framework.

9.1.4 Core Components of the Protocol

1. Advance Informed Agreement (AIA) Procedure

A fundamental feature of the Protocol, the AIA mechanism ensures that importing countries receive prior notification and make informed decisions before accepting first-time shipments of LMOs intended for environmental release.

Steps

- Exporting party submits detailed LMO dossier.
- Importing party conducts **risk assessment**.
- Decision communicated within 270 days.

2. Biosafety Clearing-House (BCH)

An online information exchange system (https://bch.cbd.int), the BCH facilitates **transparent communication** among parties regarding:

- National biosafety laws.
- LMO approvals.
- Risk assessment reports.
- Decisions under AIA.

3. Risk Assessment and Risk Management

The Protocol mandates **scientifically sound risk assessments** to evaluate:

- Direct and indirect environmental effects.
- Potential gene flow to wild relatives.

- Effects on non-target organisms.
- Human health concerns.

It promotes **risk management plans** to mitigate identified risks throughout the LMO lifecycle.

4. Handling, Transport, Packaging, and Identification (HTPI)

CPB requires clear **labeling and documentation** of LMOs to ensure traceability during movement. For LMOs-FFP, the phrase **"may contain LMOs"** is a minimum requirement.

5. Socio-economic Considerations

Parties may consider **socio-economic impacts**, especially regarding the **traditional knowledge and practices of indigenous and local communities** in biosafety decision-making (Article 26).

6. Liability and Redress (Nagoya–Kuala Lumpur Supplementary Protocol)

This supplementary instrument (2010) addresses **legal liability for damages** caused by LMOs, including ecological and socio-economic harm.

9.1.5 Implementation Mechanisms

1. National Biosafety Frameworks (NBFs)

Each Party is expected to develop an NBF comprising:

- A **regulatory regime** (laws, guidelines).
- An **institutional framework** (competent authorities, advisory bodies).
- **Risk assessment capacity**.
- **Public participation mechanisms**.

India, for instance, has implemented the **Environment Protection Act (EPA), 1986**, and established authorities like the **Genetic Engineering Appraisal Committee (GEAC)** to operationalize the CPB.

2. Capacity Building

Developing countries receive **technical and financial assistance** to enhance biosafety expertise through:

- UNEP-GEF projects.
- Regional cooperation networks.
- Training programs for regulators and scientists.

3. Public Participation and Transparency

The Protocol promotes **informed public involvement**, particularly in decisions involving LMO approvals, risk assessments, and release procedures.

9.1.6 Significance in the Indian Context

India ratified the Cartagena Protocol in **2003**, aligning its biosafety protocols with CPB mandates. Its implementation led to:

- Framing of the **Rules for the Manufacture, Use, Import, Export and Storage of Hazardous Microorganisms/Genetically Engineered Organisms or Cells (1989)**.
- Institutionalization of bodies such as:
  - **RDAC** – Recombinant DNA Advisory Committee
  - **IBSC** – Institutional Biosafety Committee
  - **RCGM** – Review Committee on Genetic Manipulation
  - **GEAC** – Genetic Engineering Appraisal Committee

India's biosafety approach is **cautious but enabling**, ensuring rigorous risk evaluation without compromising on biotechnology advancement.

9.1.7 Challenges and Future Directions

Challenges	**Strategic Directions**
Capacity limitations in LMO risk evaluation	More investment in training and risk modeling tools
Divergences in national regulatory systems	Harmonization of biosafety protocols across borders
Rapid innovation outpacing regulation	Dynamic updating of risk assessment guidelines
Traceability and detection of LMOs	Development of molecular markers and digital tracking systems

9.1.8 Conclusion

The Cartagena Protocol on Biosafety stands as a **cornerstone of international biosafety governance**, bridging the demands of biotechnological innovation with ecological responsibility. Its emphasis on **risk assessment, precautionary action, transparency, and public participation** makes it a globally respected model. For nations like India, its successful implementation is vital for ensuring that **modern biotechnology progresses without compromising environmental integrity or human health**.

9.2 Biosafety Regulations in India: Guidelines and Regulatory Authorities

9.2.1 Introduction

India, as one of the earliest adopters of biotechnology in agriculture and health, recognized the importance of biosafety early in its policy framework. To ensure that **genetically modified organisms (GMOs)** and **products of modern biotechnology** are developed, tested, and deployed safely, India has established a **comprehensive legal and regulatory framework**. This framework, rooted in environmental legislation, is supported by **guidelines, protocols, and institutional mechanisms** that govern the safe use of genetically engineered organisms (GEOs) in research, development, testing, production, and commercialization.

9.2.2 Legal Foundation: Environment (Protection) Act, 1986

The central legal instrument underpinning biosafety in India is the:

Environment (Protection) Act, 1986 (EPA)

Under this Act, the **Rules for the Manufacture, Use, Import, Export and Storage of Hazardous Microorganisms/Genetically Engineered Organisms or Cells, 1989**—commonly known as **Rules, 1989**—were framed by the **Ministry of Environment, Forest and Climate Change (MoEFCC)** in consultation with the **Department of Biotechnology (DBT)**.

These Rules provide a **statutory basis for regulating biotechnology** and form the backbone of India's biosafety regime.

9.2.3 Key Objectives of the Biosafety Regulations

- To prevent **adverse impacts of GMOs/LMOs** on the environment, biodiversity, and human/animal health.
- To regulate and oversee **research, testing, import, and commercial release** of genetically engineered products.
- To ensure **institutional and procedural transparency** in the risk assessment and approval process.
- To promote **safe innovation** in biotechnology through scientifically informed guidelines.

9.2.4 Regulatory Authorities and Their Functions

India's biosafety regulatory system involves **six competent authorities** established under the Rules, 1989, each with specific roles in risk evaluation and decision-making.

1. **Recombinant DNA Advisory Committee (RDAC)**
   - **Constituted by**: Department of Biotechnology (DBT)
   - **Mandate**: Advises on the development of recombinant DNA (rDNA) technology, biosafety policy formulation, and ethical considerations.
   - **Function**: Scientific foresight and periodic review of global biosafety trends and domestic preparedness.
2. **Institutional Biosafety Committee (IBSC)**
   - **Constituted at**: Institution level (universities, research labs, industries)
   - **Composition**: Head of the institution, biosafety officers, scientists, and a nominee from DBT.
   - **Function**
     - Approves contained research with GMOs.
     - Conducts **initial risk assessment** and containment strategy review.
     - Reports incidents and compliance to RCGM.
3. **Review Committee on Genetic Manipulation (RCGM)**
   - **Constituted by**: DBT
   - **Composition**: Scientific experts, ministry representatives, and legal/ethical advisors.
   - **Function**
     - Approves **small-scale field trials** and transgenic research.
     - Evaluates **preclinical biosafety data**.
     - Develops **protocols and biosafety guidelines**.
     - Oversees IBSCs and ensures containment infrastructure.
4. **Genetic Engineering Appraisal Committee (GEAC)**
   - **Constituted by**: MoEFCC
   - **Highest decision-making authority** under the Rules, 1989.
   - **Function**:
     - Approves **large-scale environmental release**, commercialization, and import/export of GMOs.
     - Conducts **environmental risk assessments (ERA)** and evaluates socio-economic implications.
     - Final authority on **product deregulation and monitoring post-release**.

5. State Biotechnology Coordination Committee (SBCC)

- **Constituted by**: State Governments
- **Function**
  - Monitors field trials and commercial GMO usage within states.
  - Coordinates with GEAC and central bodies for enforcement.

6. District Level Committee (DLC)

- **Constituted by**: District administration
- **Function**
  - Ensures local level monitoring of field trials.
  - Investigates complaints and reports violations to SBCC/GEAC.

9.2.5 Guidelines and Protocols

India has developed a series of **biosafety guidelines** and **standard operating procedures (SOPs)** to standardize GMO handling and risk evaluation. Key among them:

- **Recombinant DNA Safety Guidelines, 1990** (DBT)
- **Revised Guidelines for Research in Transgenic Plants and Containment, 1998**
- **Guidelines for the Safety Assessment of Foods Derived from GE Plants, 2008**
- **Protocols for Confined Field Trials of GE Plants, 2008**
- **Regulations and Guidelines on Biosafety of Recombinant DNA Research and Biocontainment, 2017**

These documents guide laboratories, industry, and researchers in **best practices for biosafety**, containment, waste disposal, monitoring, and documentation.

9.2.6 Monitoring, Compliance, and Enforcement

- **Monitoring Committees**: Project-wise monitoring teams are assigned during field trials.
- **Post-release surveillance**: Mandatory for approved transgenic crops to detect unintended consequences.
- **Penalties and liabilities**: Non-compliance with Rules, 1989, can lead to **legal prosecution, cancellation of approvals**, and **environmental damage liabilities** under EPA provisions.

9.2.7 India's Approach: Precautionary Yet Progressive

India follows a **science-based yet precautionary approach** to biosafety, ensuring that:

- **Scientific rigor**, ethical scrutiny, and **transparency** underpin the regulatory process.
- Public participation, including stakeholder consultations and right-to-know mechanisms, are encouraged.
- **Capacity-building programs** through DBT and MoEFCC strengthen institutional and regional biosafety infrastructure.

9.2.8 Recent Developments

- **Genome Editing Guidelines (2022)**: DBT notified new guidelines distinguishing **site-directed nuclease (SDN)-1 and SDN-2 products** from transgenics for regulatory relaxation.
- **Digital platforms**: Implementation of the **Biosafety Research Level (BRL) tracking system** and online portals for GMO application and tracking.
- **Alignment with international standards**: India's biosafety norms are increasingly harmonized with the **Cartagena Protocol on Biosafety** and **OECD guidelines**.

9.2.9 Conclusion

India's biosafety regulatory architecture is **multilayered, participatory, and evolving**—anchored in robust legal provisions and proactive institutional oversight. By balancing **scientific innovation and ecological prudence**, the country has positioned itself as a **global leader in responsible biotechnology governance**. Continued refinement, stakeholder engagement, and transparency will be crucial as India navigates **emerging biotechnologies such as synthetic biology, gene drives, and genome editing**.

9.3 International Biosafety Policies and Comparison

9.3.1 Introduction

With the global expansion of biotechnology, particularly **genetically modified organisms (GMOs)** and **synthetic biology applications**, the need for robust and harmonized **biosafety frameworks** across nations has become paramount. While countries share a common objective—**to ensure the safe use of biotechnology**—their regulatory philosophies, implementation strategies, and legal interpretations **diverge significantly**. This section critically examines **key international biosafety policies** and presents a comparative evaluation of prominent biosafety regimes, with particular reference to **India's position** in the global landscape.

9.3.2 The Global Foundation: Cartagena Protocol on Biosafety (2000)

The **Cartagena Protocol**, adopted under the Convention on Biological Diversity (CBD), is the first international legal instrument **dedicated exclusively to biosafety**. It regulates **transboundary movement, handling, and use of living modified organisms (LMOs)** with a potential risk to biodiversity and human health.

Key Elements

- **Advanced Informed Agreement (AIA)** for first-time import of LMOs.
- **Risk assessment and risk management** obligations.
- **Public awareness and participation.**
- **Biosafety Clearing-House (BCH)** for information exchange.

As of 2025, **173 countries** are parties to the Protocol, including India.

India aligns its national policies (e.g., Rules, 1989; GEAC guidelines) in compliance with Cartagena obligations, ensuring **prior risk assessment**, public notification, and **decision-making transparency** for LMO import/export.

9.3.3 National Biosafety Frameworks: Comparative Perspectives

Let us now analyze and compare the biosafety policies of selected key countries and blocs: **United States, European Union, India, and Brazil**—based on regulatory philosophy, risk assessment methodology, institutional framework, and public engagement.

A. United States: Product-Based and Deregulatory

- **Regulatory Approach**: Primarily **product-based**, focusing on the **end characteristics** of the organism rather than the process (e.g., genetic engineering).
- **Key Agencies**
  - **USDA-APHIS** (Animal and Plant Health Inspection Service)
  - **EPA** (Environmental Protection Agency)
  - **FDA** (Food and Drug Administration)
- **Unique Features**
  - Streamlined approval for **gene-edited crops** (e.g., CRISPR-derived products).
  - Strong **scientific evidence-driven risk assessment**.
  - Voluntary pre-market consultations.

The U.S. does not enforce a single biosafety law but relies on the **Coordinated Framework for Regulation of Biotechnology (1986, revised 2017)**.

B. European Union: Precautionary and Process-Based

- **Regulatory Approach**: Strongly **precautionary** and **process-based,** emphasizing the method of genetic modification.
- **Legislative Instruments**:
  - **Directive 2001/18/EC** on the deliberate release of GMOs
  - **Regulation (EC) No 1829/2003** on GM food and feed
- **Oversight Body**: **European Food Safety Authority (EFSA)**
- **Key Features**
  - **Mandatory labeling and traceability** of GM products.
  - **Public participation and ethical review** as essential components.
  - Restrictions on field trials and commercialization in several member states due to socio-political resistance.

EU's approach reflects a high level of **consumer and environmental protection**, often resulting in **lengthy and cautious approval timelines**.

C. India: Precautionary and Layered Institutional Mechanism

- **Regulatory Approach**: **Precautionary, science-based, and hierarchical**.
- **Legal Basis**: **EPA, 1986 and Rules, 1989**
- **Major Regulatory Bodies**: **GEAC, RCGM, IBSC**
- **Recent Trends**:
  - Emphasis on **contained field trials**, monitoring, and **post-release biosurveillance**.
  - 2022 guidelines allow **exemption of SDN-1/SDN-2 genome editing** from GEAC clearance.

India's biosafety regime is increasingly aligning with **international best practices**, while retaining autonomy to address **domestic socio-political and agro-ecological contexts**.

D. Brazil: Progressive and Risk-Based

- **Regulatory Body**: **National Biosafety Technical Commission (CTNBio)**
- **Approach**: Risk-based, science-informed with streamlined approval.
- **Notable Policies**

  - Fast-track approval of **gene-edited organisms**.
  - Integration of **environmental, health, and socio-economic** considerations.
- **Public Engagement**: Mandatory public consultations for commercial releases.

Brazil represents a **pro-developmental model** that balances **economic competitiveness in agri-biotech** with regulatory oversight.

9.3.4 Key Parameters of Comparison

Parameters	United States	European Union	India	Brazil
Regulatory Focus	Product-based	Process-based	Mixed (Process + Risk)	Risk-based
Precautionary Principle	Limited	Strong	Moderate	Balanced
Central Law	None (policy framework)	Specific GMO laws	EPA & Rules, 1989	Biosafety Law (2005)
Approval Speed	Fast	Slow	Moderate	Fast
Public Participation	Voluntary	Mandatory	Emerging	Mandatory
Genome Editing Regulation	Deregulated for SDN-1	Regulated as GMO	Deregulated (SDN-1/2)	Deregulated (SDN-1/2)

9.3.5 Global Challenges in Harmonizing Biosafety

Despite common goals, harmonizing international biosafety remains a **complex challenge** due to:

- **Diverse legal systems and risk perception**.
- Varied **public attitudes toward GMOs**.
- **Trade conflicts** arising from labeling and equivalency.
- **Lack of consensus on genome editing regulation**.

Efforts by international platforms like the **OECD, FAO, WHO**, and the **Cartagena Protocol** aim to **bridge regulatory asymmetries** by promoting **mutual information exchange, capacity building**, and **science-based policy development**.

9.3.6 Conclusion

International biosafety policies exhibit **a continuum between precaution and promotion**, with countries balancing **scientific innovation, ecological protection, and public accountability**. India's biosafety regime, rooted in the **Cartagena Protocol**, continues to evolve with **greater sophistication and adaptability**, increasingly mirroring global trends in **risk assessment,**

regulatory streamlining, and **public engagement**. A comparative understanding of these frameworks equips stakeholders to participate more effectively in global **biotech governance, trade negotiations, and biosafety standardization**.

9.4 Regulatory Laboratory Practices and Institutional Biosafety Committees (IBSC)

9.4.1 Introduction

In the realm of modern biotechnology, laboratory research involving genetically modified organisms (GMOs), recombinant DNA, and other potentially hazardous biological materials demands a highly structured and regulated framework. This ensures safety for researchers, the environment, and the general public. Regulatory Laboratory Practices and Institutional Biosafety Committees (IBSCs) play pivotal roles in implementing, supervising, and maintaining biosafety at the institutional level in compliance with national and international guidelines.

9.4.2 Concept and Importance of Regulatory Laboratory Practices

Regulatory Laboratory Practices are a set of codified biosafety principles and operational standards designed to control potential risks associated with biotechnology research, especially those involving GMOs, transgenic organisms, or pathogenic agents. These practices ensure:

- Containment of biohazards within the laboratory environment.
- Safe handling, storage, and disposal of recombinant materials.
- Prevention of unintentional release or exposure.
- Documentation and traceability of genetic constructs and biological agents.

These practices are not just internal protocols but are legally enforceable and monitored under national biosafety regulations. In India, such practices are governed primarily under the **"Rules, 1989"** of the Environment Protection Act, 1986, which relate to the manufacture, use, import, export, and storage of hazardous microorganisms/genetically engineered organisms or cells.

9.4.3 Institutional Biosafety Committees (IBSCs): Structure and Function

The **Institutional Biosafety Committee (IBSC)** is a mandatory regulatory body in every Indian institution engaged in biotechnology research. These committees serve as the **first level of review** and monitoring of biosafety and biohazard-related practices within the institution.

A. Formation of IBSC

As per the Department of Biotechnology (DBT), Ministry of Science and Technology, Government of India:

- An IBSC must be established in all institutions carrying out recombinant DNA work.
- It must be **registered with DBT** before commencing any such work.

B. Composition of IBSC

A standard IBSC typically includes

1. **Head of the Institution (Chairperson)** – Senior-most authority, overseeing compliance.
2. **Three or more internal scientists/experts** – With relevant expertise in molecular biology, microbiology, or genetic engineering.
3. **Medical Officer or Biosafety Officer** – To assess and advise on health-related hazards.
4. **One external expert (nominated by DBT)** – Serving as the **DBT nominee** to ensure impartiality and compliance with national standards.
5. **Member Secretary** – Usually a senior faculty responsible for documentation, meeting coordination, and liaison with DBT.

C. Major Functions of IBSC

- **Risk Assessment and Containment Classification:** Evaluate proposed experiments and categorize them based on biosafety levels (BSL).
- **Review and Approve Protocols:** Approve research involving recombinant DNA, gene editing, or other high-risk activities.
- **Supervise Laboratory Practices:** Ensure that laboratories follow Good Laboratory Practices (GLP) and Biosafety Levels (BSL) appropriate to the risk.
- **Training and Awareness:** Organize training sessions, safety drills, and biosafety awareness among researchers and staff.
- **Documentation and Reporting:** Maintain minutes of meetings, compliance reports, incident reports, and annual biosafety statements.
- **Communication with RCGM:** Serve as the liaison with the **Review Committee on Genetic Manipulation (RCGM)** and other apex regulatory bodies.

9.4.4 IBSC in National Biosafety Governance Framework

The IBSC is an **institution-level regulator** integrated within a multi-tiered national biosafety governance structure. Its decisions and reports feed into higher regulatory authorities such as:

- **RCGM** under DBT, which oversees research-level biosafety issues.
- **GEAC (Genetic Engineering Appraisal Committee)** under the Ministry of Environment, Forest and Climate Change (MoEFCC), which handles environmental release and commercialization aspects.

IBSCs play a central role in

- Early detection of regulatory non-compliance.
- Preventing biosafety incidents.
- Strengthening institutional accountability.
- Facilitating seamless integration between research and regulation.

9.4.5 Challenges and Way Forward

Challenges

- **Variability in Institutional Compliance:** Some research institutions lack well-trained personnel or infrastructure to implement strict biosafety measures.
- **Inadequate Training and Awareness:** Researchers may be unaware of updated biosafety norms or underestimate risks.
- **Under-reporting of Incidents:** Fear of administrative backlash may lead to suppression of minor safety breaches.
- **Lack of Regular Monitoring:** Absence of third-party auditing can lead to laxity in biosafety practice adherence.

Recommendations

- Periodic **auditing of IBSCs** by central authorities.
- Institutional **capacity-building** through advanced training.
- Ensuring **transparency and public accountability** in biosafety oversight.
- Implementation of **digital record-keeping** and **online portals** for biosafety approvals and monitoring.

9.4.6 Conclusion

Institutional Biosafety Committees (IBSCs), along with robust regulatory laboratory practices, form the backbone of biosafety infrastructure in biotechnology research institutions. Their presence ensures that research on GMOs and hazardous agents is conducted responsibly, minimizing risks to human health, biodiversity, and the environment. In an era where scientific advancement must walk hand-in-hand with societal accountability, IBSCs play a pivotal role in bridging this gap and fostering ethical innovation.

PART-III
Risk and Safety Assessment of Transgenics

10

Transgenic Patents Environmental and Food Safety Concerns

10.1 Public Concerns About GMOs

10.1.1 Introduction

Genetically Modified Organisms (GMOs), particularly transgenic crops, represent one of the most transformative breakthroughs in modern biotechnology. Despite their promise to address global challenges such as food insecurity, pest infestations, and climate resilience, GMOs have also triggered significant public concern. These concerns, though varying across regions and socio-political contexts, typically revolve around **human health**, **environmental safety**, **ethical considerations**, **corporate control**, and **regulatory transparency**. Understanding these public apprehensions is crucial for biosafety policymaking and responsible biotechnology governance.

10.1.2 Human Health-Related Concerns

A. Allergenicity and Toxicity

One of the most widely voiced apprehensions is that transgenic foods might introduce **new allergens** or **toxic compounds** into the food chain. The insertion of foreign genes, particularly from non-food species (e.g., bacteria, viruses), raises the possibility that novel proteins could elicit allergic responses or metabolic imbalances in humans.

Example: The case of a transgenic soybean incorporating a Brazil nut gene was withdrawn from commercialization after tests indicated it could cause allergic reactions in individuals sensitive to tree nuts.

B. Antibiotic Resistance Markers

Many GMOs use **antibiotic resistance genes** as selectable markers. Concerns persist that such genes could be horizontally transferred to human gut microflora, potentially exacerbating the global issue of **antibiotic resistance**.

C. Long-term Health Impacts

There is a perceived lack of **longitudinal epidemiological data** to assess the chronic effects of GMO consumption. Critics argue that without comprehensive long-term feeding studies, safety claims remain incomplete or speculative.

10.1.3 Environmental and Ecological Concerns

A. Gene Flow and Biodiversity Erosion

One of the most cited risks is **gene escape**—the transfer of transgenes to wild relatives or non-GM crops via pollen dispersal. This can lead to **unintended hybridization**, threatening local plant diversity and ecosystem stability.

Example: In Mexico, a center of origin for maize, gene flow from GM maize into traditional landraces raised alarms about biodiversity loss.

B. Emergence of Superweeds and Pest Resistance

The continuous cultivation of herbicide-tolerant or insect-resistant GM crops may lead to the **evolution of resistant weed species or insect pests**, demanding even higher chemical inputs and undermining the benefits of GM technologies.

C. Impact on Non-target Organisms

GM crops producing toxins like **Bt (Bacillus thuringiensis)** may unintentionally affect **non-target organisms** such as pollinators, soil microbes, or beneficial insects.

Controversy: Early laboratory studies suggested that Bt maize might affect Monarch butterfly larvae feeding on milkweed near GM fields, though subsequent field studies provided nuanced conclusions.

10.1.4 Ethical and Socio-economic Concerns

A. Corporate Control and Farmers' Rights

Public unease also stems from **corporate monopolization of seed technology**. Patents on GMOs may restrict farmers from saving or exchanging seeds, thereby increasing their dependency on large agro-biotech companies.

In India, widespread protests erupted against Bt cotton royalties, sparking debates on seed sovereignty and pricing control.

B. Labeling and Consumer Choice

Many consumers advocate for the **mandatory labeling** of GMO foods to make informed dietary choices. Lack of labeling is often viewed as a violation of autonomy and food ethics.

C. Religious and Cultural Sensitivities

Some GMOs, especially those involving animal genes inserted into plant systems, may **conflict with religious or cultural dietary codes**, raising moral objections among certain communities.

10.1.5 Trust Deficit and Regulatory Transparency

The public's perception of GMOs is heavily influenced by the **credibility of regulatory institutions**. In many cases, there is:

- A **lack of public engagement** in GMO approval processes.
- Concerns about **industry influence** on regulatory decisions.
- Deficiencies in **independent, peer-reviewed biosafety assessments**.

The result is a pervasive **trust deficit**, particularly in countries where civil society involvement in biosafety governance is limited.

10.1.6 Addressing Public Concerns: A Multi-Stakeholder Imperative

To bridge the gap between scientific advances and public acceptance, several measures must be implemented:

1. **Transparent Risk Communication**: Simplifying scientific information and biosafety assessments for the lay public.
2. **Robust Regulatory Oversight**: Ensuring GMO approval is based on independent, multidisciplinary evaluations.
3. **Participatory Policy-Making**: Involving farmers, consumers, and civil society in biosafety discussions.
4. **Ethical Technology Deployment**: Balancing innovation with moral and cultural sensitivities.
5. **Labeling and Right to Choose**: Mandating clear labeling and traceability for all GM-derived products.

10.1.7 Conclusion

Public concerns about GMOs are multifaceted, reflecting a complex interplay of scientific, ethical, socio-political, and cultural factors. While the biotechnology community emphasizes evidence-based risk assessments, public apprehensions stem from both real and perceived risks, which cannot be dismissed as irrational. A sustainable and inclusive GMO policy must recognize these concerns and work toward **transparency, accountability, ethical responsibility, and participatory governance**. Only then can the promises of biotechnology be fully realized in harmony with public interest.

10.2 Biosafety Aspects – Invasiveness, Weediness, and Gene Flow

10.2.1 Introduction

Genetically Modified Organisms (GMOs), especially transgenic plants, are introduced into agricultural ecosystems with the intent of enhancing productivity, stress tolerance, or pest resistance. However, the release of such organisms into open environments raises significant biosafety concerns regarding their potential **invasiveness**, propensity for **weediness**, and capacity for **gene flow** to wild or cultivated relatives. These characteristics can lead to ecological imbalances, biodiversity loss, and unintended agronomic consequences if not properly assessed and managed.

10.2.2 Invasiveness of Transgenic Plants

A. Definition and Concept

Invasiveness refers to the ability of a species, particularly non-native or engineered organisms, to establish, proliferate, and spread aggressively in ecosystems where they are introduced, often outcompeting native flora and disrupting ecological equilibrium.

B. Potential of GMOs to Become Invasive

Although most crops have undergone domestication processes that reduce their fitness in the wild, genetic modification can inadvertently **restore or enhance fitness traits**, enabling certain transgenic plants to behave like invasive species.

Example: A transgenic crop with improved abiotic stress tolerance (e.g., drought or salinity resistance) might survive and proliferate in marginal or natural habitats, leading to unintended colonization.

C. Environmental Consequences

- **Suppression of native species** and disruption of natural plant communities.
- **Altered nutrient cycling** and habitat structures.
- **Unpredictable interactions** with insects, soil microbes, and animals.

D. Assessment and Mitigation

Pre-release environmental assessments are mandated to evaluate the **invasiveness potential** using ecological modeling, trait analysis, and field trials. Mitigation strategies include **biological confinement** and **post-release monitoring**.

10.2.3 Weediness in GM Crops

A. Weediness: Definition and Criteria

Weediness is the tendency of a plant to persist and propagate aggressively in agricultural or disturbed habitats, often without cultivation. A plant is considered "weedy" if it competes with crops, reduces yields, or complicates field management.

B. Risk from Transgenic Traits

Transgenic modifications may inadvertently enhance **traits associated with weediness**, such as:

- **Increased seed dormancy**
- **Rapid vegetative growth**
- **Resistance to pests, herbicides, or harsh conditions**

Case Insight: Herbicide-resistant GM crops (e.g., glyphosate-tolerant soybean or canola) can potentially become volunteer weeds in subsequent planting seasons, complicating weed control strategies.

C. Implications

- Escaped transgenic volunteers may require **more intensive chemical control**, reversing the environmental benefits of GMO adoption.
- **Weed–crop hybrids** can emerge where gene flow occurs with related weedy species.

10.2.4 Gene Flow and Transgene Escape

A. Definition and Pathways

Gene flow refers to the movement of genetic material between populations through **pollen dispersal**, **seed migration**, or **vegetative propagules**. In the context of GMOs, **transgene flow** is the unintentional transmission of engineered genes to:

- Wild relatives (introgression)
- Non-GM crops (adventitious presence)
- Weedy species

B. Risk Factors Influencing Gene Flow

- **Sexual compatibility** with wild relatives.
- **Pollination biology** (e.g., insect vs. wind pollination).
- **Proximity to non-GM or wild populations**.
- **Persistence of pollen and seeds** in the environment.

C. Ecological Implications

- Creation of **transgenic wild hybrids** with novel traits (e.g., herbicide resistance).
- Undermining of **biodiversity conservation**, especially in centers of origin.
- Contamination of **organic and traditional crop systems**.

D. Socio-economic Consequences

- **Legal and trade disputes** over unintentional presence of patented genes.
- Loss of **market access** for GMO-sensitive countries (e.g., EU bans).
- Challenges to **seed sovereignty** and intellectual property enforcement.

E. Strategies for Gene Flow Management

1. **Temporal Isolation** – Staggered flowering periods between GM and non-GM crops.
2. **Spatial Isolation** – Buffer zones and separation distances.
3. **Biological Confinement** – Male sterility, chloroplast transformation (maternal inheritance), gene use restriction technologies (GURTs).
4. **Physical Barriers** – Greenhouses, screen houses for confined field trials.

10.2.5 Regulatory and Risk Assessment Approaches

International biosafety guidelines, such as those under the **Cartagena Protocol on Biosafety**, mandate **comprehensive risk assessments** before environmental release. Key components include:

- Trait-specific ecological impact studies.
- Gene flow modeling using probabilistic frameworks.
- Monitoring protocols for **post-release ecological dynamics**.

In India, regulatory bodies such as the **Genetic Engineering Appraisal Committee (GEAC)** and **Review Committee on Genetic Manipulation (RCGM)** oversee these assessments for environmental release approvals.

10.2.6 Conclusion

The concerns of **invasiveness, weediness, and gene flow** form the core of biosafety evaluations for transgenic organisms. These phenomena are intricately linked with ecological sustainability, agricultural stability, and long-term societal impacts. Ensuring that GM crops do not disrupt natural or managed ecosystems requires rigorous scientific assessment, robust containment strategies, and transparent regulation. A **precautionary yet progressive biosafety framework** must be embraced, integrating cutting-edge biotechnology with ecological stewardship.

10.3 Gene Transfer and Impacts on Non-Target Organisms

10.3.1 Introduction

The environmental release of genetically modified organisms (GMOs), particularly transgenic crops, has prompted widespread biosafety assessments to understand the consequences of gene transfer and their cascading ecological effects. One critical area of concern is the **unintended gene flow to non-target organisms** and the resulting **impacts on biodiversity, ecosystem stability, and food webs**. This section evaluates the biological routes of gene transfer and their potential risks to non-target organisms, drawing on both empirical studies and precautionary biosafety principles.

10.3.2 Mechanisms of Gene Transfer in GMOs

Gene transfer refers to the movement of genetic material from transgenic organisms to other biological entities through natural or anthropogenic processes. Two primary types of gene transfer are considered in the context of biosafety:

A. Vertical Gene Transfer (VGT)

- Occurs through **sexual reproduction** between GM crops and compatible wild or weedy relatives.
- Leads to the incorporation of transgenes into the **germline of progeny**, potentially resulting in **persistent heritable changes**.

B. Horizontal Gene Transfer (HGT)

- Involves the transfer of genetic material **across species boundaries**, typically between transgenic plants and microorganisms (e.g., soil bacteria, fungi).
- Though rare, HGT raises concerns over **novel trait dissemination**, such as antibiotic resistance marker genes being transferred to **pathogenic microbes**.

10.3.3 Impacts on Non-Target Organisms (NTOs)

Transgenic crops expressing novel traits (e.g., insecticidal proteins, herbicide tolerance, viral resistance) can influence organisms **not intended to be affected** by the genetic modification. These non-target organisms include **beneficial insects, soil biota, birds, mammals, aquatic organisms**, and **symbiotic microbes**. The impacts may be direct or indirect, immediate or delayed.

A. Insecticidal GM Crops (e.g., Bt Crops)

Bt (Bacillus thuringiensis) crops express crystal (Cry) proteins toxic to specific insect pests. However, concerns have emerged regarding their potential impacts on:

- **Pollinators (e.g., honeybees)** – Exposure to Bt pollen or nectar may affect foraging behavior or larval development.
- **Predatory insects and parasitoids** – These beneficial insects may be harmed if they consume prey containing Bt toxins.
- **Non-target Lepidoptera** – Monarch butterflies (Danausplexippus), for instance, may be affected by Bt maize pollen deposited on milkweed.

Scientific Insight: A study by Losey et al. (1999) indicated that monarch butterfly larvae feeding on milkweed dusted with Bt corn pollen had higher mortality rates under laboratory conditions, although subsequent field assessments suggested more nuanced effects.

B. Soil Ecosystems and Microbial Communities

Transgenic crops may alter the **rhizosphere environment**, affecting:

- **Nitrogen-fixing bacteria** and **mycorrhizal fungi**
- **Decomposition rates and nutrient cycling**
- **Soil enzymatic activity and microbial diversity**

Transgenic residues (roots, leaves) with modified biochemical compositions can impact **soil detritivores and microbial decomposers**, altering **ecosystem services**.

C. Aquatic Organisms

Genes or gene products (e.g., Bt toxins) may leach into waterways through **agricultural runoff**, affecting:

- **Freshwater invertebrates**, such as Daphnia
- **Amphibians**, which are highly sensitive to chemical alterations in their habitats
- **Algal and planktonic communities**, which form the base of aquatic food chains

D. Birds and Mammals

Indirect effects on **higher trophic levels** may result from:

- Reduced prey populations due to toxicity
- Altered food quality or availability
- Ingestion of transgene-expressing seeds or plant parts

Although acute toxicity to vertebrates is often low, **long-term feeding studies** are necessary to rule out subtle physiological or reproductive impacts.

10.3.4 Ecological Concerns Arising from Gene Transfer

- **Trophic cascades:** Disruption in predator–prey dynamics may lead to ecological imbalances.
- **Resistance development:** Non-target pests may evolve resistance to transgene products, creating **secondary pest outbreaks**.
- **Biodiversity erosion:** Altered interactions among organisms may lead to declines in native species.
- **Loss of ecosystem services:** Declines in pollinators, decomposers, or natural pest regulators can impair key ecosystem functions.

10.3.5 Risk Assessment Approaches

Rigorous risk assessment protocols are crucial before field release of transgenic organisms. These include:

A. Tiered Testing Framework

1. **Laboratory studies** – Evaluate acute and chronic toxicity on indicator species.
2. **Greenhouse experiments** – Simulate semi-controlled interactions.
3. **Field trials** – Examine ecosystem-level impacts in real environments.

B. Selection of Surrogate Species

Testing is done on **ecologically relevant non-target species** that serve as representatives of functional groups (pollinators, predators, decomposers).

C. Molecular Characterization

Assessment of **gene expression stability**, **promoter behavior**, and **target specificity** of transgene products.

D. Environmental Monitoring

Post-release monitoring under regulatory supervision to detect any **unintended effects on non-target species**, often mandated under **biosafety protocols** like the Cartagena Protocol.

10.3.6 Regulatory Safeguards and Mitigation Strategies

- **Gene confinement technologies** to limit exposure to non-target organisms (e.g., plastid transformation, male sterility).
- **Refuge strategies** to delay resistance and maintain non-target diversity.

- **Labeling and traceability** for effective monitoring and management.
- **Adaptive management** involving ongoing research and regulatory revisions.

10.3.7 Conclusion

Gene transfer from GMOs to non-target organisms, and the subsequent ecological interactions, present a **complex biosafety challenge** in biotechnology. A science-based, precautionary approach is essential for risk management, guided by thorough ecological understanding and robust testing protocols. Safeguarding **non-target organisms**, which play vital roles in ecosystem health and agricultural sustainability, must remain a cornerstone of **transgenic risk governance**.

11

Risk Assessment and Safety Evaluation of GM Crops

11.1 Principles and Framework for Risk Assessment

11.1.1 Introduction

Risk assessment is an indispensable scientific tool in evaluating the safety and potential impacts of genetically modified (GM) crops on human health, animal welfare, and the environment. It provides a structured, evidence-based framework for identifying hazards, characterizing risks, and proposing mitigation strategies before the approval and commercialization of GMOs. In the context of biotechnology and regulatory biosafety, risk assessment ensures that GM crops meet both **national and international standards for biosafety and sustainability**.

11.1.2 Definition of Risk Assessment

Risk assessment is defined as the **systematic process of identifying, evaluating, and characterizing the potential adverse effects** arising from exposure to a GMO or its products, under specified conditions of use.

Risk = Hazard × Exposure

Where hazard is the potential to cause harm, and exposure is the likelihood of encountering the hazard.

11.1.3 Objectives of Risk Assessment in GM Crops

- To evaluate the **safety of GM crops** for human and animal consumption.
- To assess the **potential environmental impacts**, including effects on biodiversity, gene flow, and ecosystem interactions.
- To ensure that **regulatory decisions are science-based**, transparent, and precautionary.
- To guide the development of **containment, mitigation, and monitoring strategies**.
- To **build public trust and regulatory confidence** through rigorous and transparent evaluation.

11.1.4 Core Principles of Risk Assessment

A. Science-Based and Evidence-Driven

Risk assessment must rely on **validated scientific methods**, empirical data, and reproducible experiments, free from bias or assumption.

B. Case-by-Case Evaluation

Each GM crop is assessed **individually**, based on its unique **transgene, host species, intended use, and release environment**. No universal conclusions are drawn across all GMOs.

C. Comparative Approach

The GM crop is compared with its **conventional counterpart**, often termed the "**non-GM comparator**," to identify any differences that may pose risks.

D. Step-Wise Process

Risk assessment follows a **tiered and progressive approach**, beginning with laboratory analyses and proceeding to confined field trials and larger environmental evaluations.

E. Transparency and Precaution

In the presence of scientific uncertainty or insufficient data, the **precautionary principle** is applied. All assumptions, methods, and results must be **fully disclosed** to regulatory and public stakeholders.

11.1.5 General Framework for Risk Assessment of GM Crops

The risk assessment process is structured into four essential components:

1. Hazard Identification

- Determining **potential sources of harm**, such as:
  - Allergenicity or toxicity of novel proteins
  - Gene transfer to non-target species
  - Weediness or invasiveness
  - Adverse effects on beneficial organisms

2. Hazard Characterization

- Quantifying the **severity and nature** of the identified hazard.
- May involve **toxicological studies, molecular profiling, compositional analysis**, and **nutritional equivalence tests**.

3. Exposure Assessment

- Estimating **likelihood, frequency, and magnitude of exposure** to the hazard.
- Considers **routes of exposure**, such as ingestion, inhalation, or environmental release.

4. Risk Characterization

- Integration of hazard and exposure data to define the **overall risk level**.
- Includes a **quantitative or qualitative description** of uncertainties and assumptions.

11.1.6 Molecular and Phenotypic Characterization of GM Crops

A critical part of risk assessment involves **detailed molecular analysis** of the transgenic construct:

- **Nature and source of the transgene**
- **Insertion site(s)** and **copy number**
- **Expression patterns** across tissues and developmental stages
- **Stability of gene expression** across generations
- **Off-target effects or unintended changes**

Simultaneously, **phenotypic assessments** are conducted to ensure agronomic traits (e.g., yield, growth habit, disease resistance) remain within acceptable boundaries compared to conventional varieties.

11.1.7 Nutritional and Toxicological Safety Assessment

- Comparative **nutritional profiling** between GM and non-GM crops is conducted to detect any changes in:
  - Macronutrients, micronutrients
  - Anti-nutrients and secondary metabolites
- **Toxicity tests** (acute, sub-chronic, and chronic) and **digestibility studies** are performed in animal models.
- **Allergenicity assessments** are carried out using in silico tools, serum screening, and pepsin digestibility tests.

11.1.8 Environmental Risk Assessment (ERA)

The ERA component addresses the **ecological implications** of cultivating GM crops in open environments:

- **Potential for gene flow** to wild or weedy relatives
- **Impact on non-target organisms**, including pollinators and natural predators

- **Development of pest or weed resistance**
- **Effect on soil biodiversity and ecosystem services**

These evaluations are performed under **confined field trials**, **multi-location testing**, and **post-release environmental monitoring**.

11.1.9 Risk Management and Mitigation

Once risks are characterized, **risk management strategies** are developed to:

- Prevent or reduce exposure (e.g., buffer zones, crop rotation)
- Employ **gene confinement mechanisms** (e.g., terminator genes, chloroplast transformation)
- Implement **Integrated Pest Management (IPM)** to delay resistance
- Conduct **post-market surveillance** and monitoring programs

11.1.10 Role of Stakeholders and Regulatory Institutions

Risk assessment must involve:

- **Developers and researchers** (responsible for generating scientific data)
- **Regulatory bodies** (for independent review and decision-making)
- **Institutional Biosafety Committees (IBSCs)** and **Review Committees on Genetic Manipulation (RCGMs)**
- **Public consultation and participation**, especially in democratic frameworks where GMO adoption is a socio-political issue

11.1.11 Conclusion

The principles and framework for risk assessment of GM crops embody a **rigorous, science-based, and precautionary approach** to ensuring biosafety. The goal is not to hinder innovation but to **enable responsible use of biotechnology** in agriculture, aligned with ecological sustainability, food safety, and public health standards. A robust risk assessment framework, integrated with transparent governance, is pivotal to fostering **scientific credibility and public trust** in genetically modified technologies.

11.2 Familiarity and Substantial Equivalence Concepts

11.2.1 Introduction

In the domain of biosafety assessment for genetically modified (GM) crops, two pivotal guiding principles—**familiarity** and **substantial equivalence**—form the scientific foundation upon which initial risk characterizations are built. These concepts facilitate a systematic and comparative approach to evaluate whether a GM crop introduces new or enhanced risks relative to its conventional counterpart. They are extensively endorsed by international bodies such as the **OECD**, **Codex Alimentarius Commission**, and national regulatory frameworks including those in India, EU, USA, and Japan.

11.2.2 Concept of Familiarity

Familiarity refers to the extent of **prior scientific knowledge, experience, and historical data** available for the host organism, the transgene, the gene product, and the target environment. It is a **contextual tool**, helping risk assessors determine whether a proposed genetic modification introduces novel elements that might require deeper investigation.

Key Dimensions of Familiarity

- **The Host Crop**: Knowledge of the biology, ecology, reproductive behavior, allergenicity, and toxicity of the non-modified crop.
- **The Transgene Source**: Genetic origin, functionality, and history of safe use.
- **The Genetic Construct**: Known behavior of the promoter, marker, and coding sequences.
- **The Receiving Environment**: Agro-climatic and ecological knowledge of the area proposed for GM crop cultivation.

Application

Familiarity is **not a substitute for risk assessment**, but an **enabling precondition** that determines:

- The **depth of data required** for specific endpoints.
- The **use of tiered testing**, where high familiarity may allow bypassing certain tests.
- **Focused risk hypotheses**, reducing unnecessary repetition of studies.

11.2.3 Concept of Substantial Equivalence

The concept of **substantial equivalence** was introduced by the **OECD in 1993** and further reinforced by the **Codex Alimentarius Commission (2003)**. It posits that **a novel food (e.g., GM crop) can be considered as safe as its conventional counterpart** if it is shown to be substantially equivalent in composition, nutrition, and safety.

Core Assumptions

- Conventional crops have a long history of safe use.
- If a GM crop does not differ significantly in its biochemical and toxicological profile, it **does not introduce additional risk**.

Comparative Parameters Include:

- **Compositional Analysis**: Proteins, fats, carbohydrates, vitamins, minerals, anti-nutrients.
- **Agronomic Traits**: Yield, maturity, pest resistance, stress tolerance.
- **Nutritional Properties**: Digestibility, bioavailability, nutrient balance.
- **Toxicological and Allergenicity Profiles**: Comparison of known allergens or toxins.

Substantial Equivalence Is Not

- A final declaration of safety
- A "yes or no" decision rule
- A rigid or unscientific assertion

Instead, it is a **starting point** for a hypothesis-driven risk assessment, helping to **highlight any differences** that require further examination.

11.2.4 Interplay Between Familiarity and Substantial Equivalence

The principles of familiarity and substantial equivalence are **interdependent**:

- Familiarity provides **contextual knowledge** and historical baselines.
- Substantial equivalence enables **comparative assessments** between the GM crop and its conventional counterpart.

If **high familiarity** is present, it may strengthen confidence in establishing substantial equivalence with **less experimental effort**. Conversely, **low familiarity** may warrant **more extensive testing**, especially in the case of novel traits (e.g., transgenic drought tolerance or nutrient biofortification).

11.2.5 Scientific and Regulatory Significance

Both concepts serve multiple purposes in **regulatory biosafety**:

- **Risk Prioritization**: Focuses scientific attention on novel or uncertain traits.
- **Streamlined Assessments**: Reduces redundancy in data requirements for well-characterized traits.
- **Facilitates International Harmonization**: Recognized across OECD countries, FAO, WHO, and Codex guidelines.
- **Enhances Public Communication**: Provides a rational basis to explain why certain GM crops are treated similarly to conventional varieties.

11.2.6 Criticisms and Evolving Interpretations

Although foundational, these concepts have not been free from criticism:

- **Subjectivity** in determining what constitutes "familiar" knowledge.
- **Insufficiency** when dealing with complex traits or multi-gene constructs.
- **Consumer distrust** when substantial equivalence is perceived as regulatory leniency.

Modern biosafety frameworks therefore treat these concepts as **tools rather than conclusions**, and emphasize **continuous refinement** through updated scientific evidence and broader public engagement.

11.2.7 Conclusion

The concepts of **familiarity** and **substantial equivalence** offer a **scientifically grounded, internationally accepted framework** for the initial screening and contextual evaluation of GM crops. While not definitive safety assurances, they provide a **rational foundation for risk assessment**, helping regulatory bodies identify areas that warrant further scrutiny or those that may proceed with standard monitoring. Together, they contribute to a **balanced and efficient biosafety system** that is both protective and enabling of responsible biotechnological innovation.

11.3 Toxicity, Allergenicity, and Environmental Impacts

11.3.1 Introduction

Genetically modified (GM) crops, engineered for traits such as insect resistance, herbicide tolerance, or nutritional enhancement, have immense potential in agriculture. However, their commercialization and deployment raise valid concerns regarding **human and animal health**, as well as **environmental sustainability**. Among these, three core areas form the cornerstone of biosafety evaluation:

1. **Toxicity** – potential to cause harm through toxic compounds;
2. **Allergenicity** – likelihood of triggering allergic responses;
3. **Environmental impacts** – unintended ecological consequences of GM crop cultivation.

Thorough evaluation in these domains ensures the safe integration of GM technology in food systems and agroecosystems.

11.3.2 Toxicity Assessment of GM Crops

Toxicity refers to the adverse physiological effects caused by toxic substances, which may be either newly expressed proteins (e.g., Bt toxins) or unintended metabolites due to genetic manipulation.

A. Types of Toxicity Evaluations

1. **Acute Toxicity Studies**
   - Typically involve a **single high-dose exposure** to laboratory animals (e.g., rats or mice) to detect any immediate adverse effect.
   - Example: Evaluation of Cry proteins (e.g., Cry1Ab) from *Bacillus thuringiensis*.
2. **Sub-chronic and Chronic Toxicity Studies**
   - Repeated dosing over an extended period (up to 90 days or more) to evaluate cumulative and long-term effects.
3. **In Vitro Toxicity Tests**
   - Employ cultured cells to evaluate cytotoxicity, genotoxicity, or metabolic disturbances.
4. **Comparative Analysis**
   - Quantitative comparison of **toxin-producing profiles** between GM crops and their non-GM counterparts to detect unintended effects.

B. Key Considerations

- Testing should include **newly expressed proteins**, **processing-induced products**, and **metabolites**.
- Use of **purified proteins** instead of plant material is common but must be structurally and functionally equivalent.
- Absence of toxicity in animal models is a prerequisite before human consumption approval.

11.3.3 Allergenicity Assessment of GM Crops

Allergenicity relates to the **potential of GM crops or their derived proteins to induce allergic responses** in sensitive individuals. Unlike toxicity, allergenicity is **not dose-dependent** and may be triggered by minute quantities.

A. Stepwise Approach for Allergenicity Evaluation

According to **Codex Alimentarius** guidelines:

1. **Source of the Gene**
   - Genes derived from known allergens (e.g., peanuts, shellfish) warrant high caution.
   - Example: The case of a Brazil nut protein transferred to soybeans (eventually halted due to allergenicity).

2. **Sequence Homology**
   - Bioinformatic comparison with known allergenic proteins (e.g., in the WHO/IUIS Allergen Database).
   - Proteins with >35% identity over 80 amino acids or identical 8-mer sequences are suspect.
3. **Pepsin Digestibility Test**
   - Allergens tend to resist pepsin digestion. Stability in **simulated gastric fluid** is a red flag.
4. **Thermal Stability**
   - Allergenic proteins are typically heat-stable and resist denaturation during cooking.
5. **Serological Testing**
   - Using sera from allergic individuals to determine **IgE binding** to the novel protein.
6. **Animal Models and In Vitro Assays**
   - Limited predictive value, but increasingly used to evaluate sensitization potential.

B. Risk Management

- Exclusion of genes from allergenic sources.
- Labeling and traceability systems to safeguard sensitive individuals.
- Post-market monitoring for adverse reactions.

11.3.4 Environmental Impacts of GM Crops

While GM crops are designed to reduce agrochemical use and improve yield, their **unintended ecological effects** require careful evaluation.

A. Gene Flow and Biodiversity

- **Horizontal gene transfer** from GM crops to wild relatives or non-GM varieties may alter **natural genetic pools**, leading to:
  - Emergence of **superweeds** (e.g., herbicide-resistant volunteers),
  - **Genetic contamination** of landraces,
  - Disruption of **conservation efforts** for native cultivars.

B. Impact on Non-target Organisms

- GM crops like Bt cotton or Bt maize produce **insecticidal proteins**, which may affect:

- **Beneficial insects** (e.g., honeybees, ladybirds),
- **Soil microbiota**, altering nitrogen fixation or decomposition processes,
- **Aquatic organisms**, through runoff of plant residues.

C. Resistance Development

- Target pests may develop **resistance** to transgenic traits, leading to pest resurgence and reduced efficacy of GM traits.
- Example: Pink bollworm resistance to Bt cotton in India and China.

D. Altered Agronomic Practices

- Dependence on herbicide-tolerant crops (e.g., glyphosate-resistant soybean) can:
  - Encourage **monocultures** and reduce crop diversity,
  - Lead to **herbicide overuse**, environmental contamination, and non-target weed resistance.

E. Ecosystem Effects

- Possible disruption of **trophic interactions**, **pollinator networks**, and **habitat dynamics**.
- Assessment includes **field trials**, **ecological modeling**, and **post-release monitoring**.

11.3.5 Integrated Risk Assessment and Monitoring

- The **ERA (Environmental Risk Assessment)** framework integrates toxicity, allergenicity, and environmental data to make regulatory decisions.
- Risk evaluation must be **science-based, transparent, and precautionary**, especially in biodiversity-rich countries like India.
- **Post-market environmental monitoring (PMEM)** ensures early detection of unintended effects.

11.3.6 Conclusion

Toxicity, allergenicity, and environmental impacts form the **triad of biosafety assessment** for GM crops. These evaluations, grounded in internationally accepted protocols and national regulations, are essential for ensuring that **genetic modifications do not compromise food safety or ecological balance**. With the rapid evolution of biotechnology, biosafety assessment frameworks must continue to be **adaptive, precautionary, and evidence-driven**, to ensure sustainable and responsible deployment of GM technologies.

12

Detection and Monitoring of Transgenics

12.1 Monitoring Strategies: Pre- and Post-Release

12.1.1 Introduction

Monitoring of genetically modified organisms (GMOs) or transgenic crops is a critical component of biosafety governance. It involves the **systematic collection of scientific data** to assess the presence, behavior, and potential impacts of transgenic organisms in controlled and natural environments. Monitoring strategies are **divided into two principal stages**:

1. **Pre-release monitoring**, which takes place during confined field trials, and
2. **Post-release monitoring**, conducted after commercial or environmental release.

These strategies ensure that transgenic technologies are introduced responsibly, maintaining **human safety**, **ecosystem integrity**, and **regulatory compliance**.

12.1.2 Objectives of Monitoring

- To evaluate the **biosafety and efficacy** of GMOs in different environments.
- To detect **unexpected adverse effects** on human health or the environment.
- To monitor **gene flow**, **resistance development**, or **weediness**.
- To ensure **compliance with biosafety regulations**, stewardship programs, and licensing agreements.
- To maintain **public trust** and provide transparency in GM deployment.

12.1.3 Pre-Release Monitoring Strategies

Pre-release monitoring is conducted in **confined field trials (CFTs)** under controlled, supervised conditions. The goal is to assess the **phenotypic behavior**, **stability of transgene expression**, and **potential ecological interactions** before environmental release.

A. Confined Field Testing Protocols

- Conducted under **institutional or national biosafety guidelines** (e.g., RCGM and GEAC in India).
- Ensures **physical, reproductive, and ecological containment** during trials.
- Trial sites are carefully selected to **avoid wild relatives**, reduce **cross-pollination risks**, and permit **monitoring of escape**.

B. Key Aspects of Pre-Release Monitoring

1. **Phenotypic Stability**
   - Observation of trait consistency (e.g., insect resistance, herbicide tolerance) over multiple generations and environments.
2. **Expression Analysis**
   - Quantification of transgene expression at molecular and protein levels using qPCR, ELISA, or immunoblotting.
3. **Environmental Interactions**
   - Assessment of **non-target organism exposure**, **pest behavior changes**, and **soil microbial interactions**.
4. **Gene Flow Analysis**
   - Monitoring pollen-mediated transfer to neighboring crops or wild species through **pollen traps**, **buffer zones**, and **molecular tracing**.
5. **Molecular Characterization and Event Identity**
   - Confirmation of **single-copy insertion**, **absence of vector backbone**, and **genetic stability** across generations.

C. Data Reporting and Regulatory Oversight

- Trial outcomes are submitted to **regulatory bodies** (e.g., GEAC in India) for scrutiny before recommending environmental release.
- **Independent third-party audits** may be mandated to ensure compliance and objectivity.

12.1.4 Post-Release Monitoring Strategies

Post-release monitoring (PRM) refers to the surveillance conducted **after transgenic crops have been approved for commercial cultivation or environmental use**. It aims to detect unintended long-term effects and ensure that the GMO behaves as expected in real-world agroecosystems.

A. Framework of Post-Release Monitoring

As defined by **Cartagena Protocol on Biosafety**, PRM includes two tiers:

1. **Case-Specific Monitoring (CSM)**
    - Targeted surveillance based on **hypothesized risks** identified during risk assessment.
    - For example, if Bt crops may affect butterflies, monitoring would focus on their populations.
2. **General Surveillance (GS)**
    - Broad, non-hypothesis-driven monitoring that captures **emerging or unforeseen effects**.
    - Includes farmer surveys, ecosystem observations, and yield trends.

B. Elements of Post-Release Monitoring

1. **Molecular Tracking of Transgene**
    - Detection of GM-specific DNA or protein markers in crops, soil, water, and associated organisms using:
        - **PCR-based methods**
        - **Lateral flow strips**
        - **ELISA or immunoassays**
        - **Next-generation sequencing (NGS)**
2. **Gene Flow and Outcrossing Monitoring**
    - Assessment of transgene introgression into **wild relatives**, **landraces**, or **non-GM fields** using:
        - **Pollen traps and seed sampling**
        - **Population genetics studies**
        - **Molecular fingerprinting**
3. **Impact on Non-Target Organisms**
    - Observation of beneficial or keystone species like **pollinators (bees, butterflies)**, **predators**, **soil fauna**, and **aquatic organisms**.
4. **Resistance Monitoring**
    - Surveillance for **evolution of resistance** in target pests or weeds, using bioassays or field sampling.
    - Example: Monitoring pink bollworm resistance to Bt cotton in India.

5. **Farmer Feedback and Socioeconomic Monitoring**
   - Collection of user experiences, agronomic performance, and market responses through participatory methods.
   - Helps assess **adoption constraints** and **unanticipated outcomes**.
6. **Monitoring Agronomic Practices**
   - Evaluation of changes in pesticide/herbicide use, soil fertility, or biodiversity due to GM adoption.
   - Supports **adaptive management strategies**.

C. Institutional Mechanisms

- Monitoring is overseen by **biosafety regulatory agencies**, often with the help of:
  - **Institutional Biosafety Committees (IBSCs)**
  - **State Biotechnology Coordination Committees (SBCCs)**
  - **District Level Committees (DLCs)**
- Monitoring reports are mandatory for **post-approval review**, **renewals**, or **revocation** of GM event licenses.

12.1.5 Integration with Risk Management

Monitoring outcomes are integrated into **risk management decisions**, such as:

- Imposing **buffer zones or isolation distances**
- Modifying **trait deployment strategies**
- Initiating **remedial action** if adverse effects are detected
- Strengthening **stewardship programs** to promote responsible use

12.1.6 Challenges and Future Directions

Despite the structured frameworks, several challenges persist:

- **Resource limitations** and lack of trained personnel in remote areas
- **Insufficient baseline ecological data**
- Need for **standardized monitoring protocols** across regions
- Incorporation of **advanced molecular tools** (e.g., CRISPR-detect, metagenomics)
- Enhancing **transparency** and **public participation** in monitoring programs

12.1.7 Conclusion

Effective pre- and post-release monitoring of transgenics is indispensable for ensuring **biosafety compliance**, **ecological harmony**, and **sustainable agricultural innovation**. A robust monitoring system, informed by science and supported by policy, fosters public trust and enables the responsible deployment of genetically engineered crops. As GM technologies evolve, so must our monitoring strategies—toward greater sensitivity, inclusivity, and anticipatory governance.

12.2 Sampling Methods and Analytical Techniques for Transgenic Detection

12.2.1 Introduction

Detection and identification of genetically modified organisms (GMOs) require robust **sampling protocols** and **analytical techniques** to ascertain the presence, quantity, and identity of transgenic components. Accurate detection is crucial for:

- **Regulatory compliance**,
- **Labeling**,
- **Risk assessment**, and
- **Trade certification**.

To ensure scientific rigor and legal admissibility, both **pre-analytical (sampling)** and **analytical** steps must be performed according to internationally validated guidelines (e.g., ISO, OECD, Codex Alimentarius, and ISTA).

12.2.2 Sampling Methods: Principles and Procedures

Sampling is the first and most critical step in the GMO detection process. An erroneous or biased sample can compromise the validity of the analysis. Hence, standardized protocols are followed to collect **representative, homogeneous, and contamination-free samples**.

A. Sampling Objectives

- To obtain a **statistically valid subset** of the whole material (seed, grain, leaf, processed food, soil, etc.)
- To ensure **uniformity and randomness** to reflect true transgene occurrence
- To minimize **cross-contamination** and **analytical bias**

B. Sampling Types

1. **Seed and Grain Sampling**
   - Performed as per **ISTA (International Seed Testing Association)** or **OECD guidelines**
   - Representative sampling from:
     - **Seed lots**
     - **Field harvests**
     - **Storage bins or silos**
2. **Plant Tissue Sampling**
   - Conducted during growth stages from **leaf, stem, flower, or pollen**
   - Used for in-field or laboratory detection of transgene expression
   - Sampling tools must be **sterile** to prevent contamination
3. **Soil and Environmental Samples**
   - Collection of **soil, water, and microbial samples** near GM field sites to monitor persistence and spread
   - Requires precise geolocation and documentation
4. **Processed Product Sampling**
   - Includes **flours, oils, and feeds**
   - Sampling must account for **homogenization levels** and **processing effects** on DNA/protein integrity

C. Sampling Protocols

- **Randomized sampling** using statistical models (e.g., hypergeometric distribution)
- Use of **sealed, labeled, and barcoded containers**
- **Chain-of-custody documentation** for traceability
- Samples must be stored under **optimal temperature and moisture conditions**

12.2.3 Analytical Techniques for GMO Detection

Detection of GMOs involves a **tiered analytical framework**, progressing from **screening to confirmation** and **quantification**.

A. DNA-Based Methods

1. **Polymerase Chain Reaction (PCR)**
   - Gold standard for GMO detection

- Detects transgene-specific DNA sequences
- Types
  - **Conventional PCR**: Qualitative detection (presence/absence)
  - **Real-Time Quantitative PCR (qPCR)**: Determines GMO content (%)
- Targets:
  - **Promoters** (e.g., CaMV35S)
  - **Selectable markers** (e.g., nptII, bar)
  - **Event-specific junction sequences**

2. **Digital PCR (dPCR)**
   - Highly sensitive and precise
   - Suitable for **low-level or trace detection**
   - Increasingly used in regulatory laboratories
3. **Loop-Mediated Isothermal Amplification (LAMP)**
   - Rapid, field-deployable alternative to PCR
   - Does not require thermal cycling
   - Useful for **on-site detection** of transgenes
4. **Next-Generation Sequencing (NGS)**
   - High-throughput detection of **known and unknown GM events**
   - Enables **whole-genome profiling**
   - Emerging in research and regulatory platforms

B. Protein-Based Methods

1. **Enzyme-Linked Immunosorbent Assay (ELISA)**
   - Detects **expressed transgenic proteins** in plant tissues or foods
   - Suitable for high-throughput screening
   - Used for Bt toxins, EPSPS, PAT, etc.
2. **Western Blotting**
   - Confirms the **identity and integrity of transgenic proteins**
   - Provides size-specific protein expression information
3. **Lateral Flow Immunoassays (LFIA)**
   - Rapid, field-friendly strip tests
   - Provide qualitative results in <10 minutes
   - Commonly used for Bt or herbicide-tolerant crops

C. Metabolite and Phenotypic Detection

1. **Targeted Metabolomics**
    - Analyzes changes in metabolites due to transgene expression
    - Used to identify **unintended effects** or **novel metabolite accumulation**
2. **Mass Spectrometry (MS) and Gas Chromatography (GC)**
    - High-precision detection of specific biochemical markers
3. **Phenotypic Bioassays**
    - Functional assays using **insect larvae (for Bt crops)** or **herbicide application** to detect trait efficacy

12.2.4 Method Validation and Quality Assurance

For results to be acceptable in regulatory or trade contexts, analytical methods must undergo strict **validation and quality control**:

- **Specificity and Sensitivity**: Ability to detect true positives and avoid false negatives
- **Limit of Detection (LOD)** and **Limit of Quantification (LOQ)**: Defined and verified
- **Reproducibility**: Across laboratories, instruments, and operators
- **Internal and External Controls**: Positive and negative controls for every run
- **Accreditation**: Laboratories must adhere to **ISO/IEC 17025** standards

12.2.5 Integration into National and International Monitoring Frameworks

- **India**: Detection and certification of GMOs is conducted by **GEAC-approved testing laboratories**, **Customs authorities**, and **export certification bodies**.
- **International bodies** like **European Food Safety Authority (EFSA)**, **USDA**, and **Codex** provide harmonized guidelines.
- Tools and results from analytical methods are critical in:
    - **Post-market surveillance**
    - **Labeling enforcement**
    - **Trade dispute resolution**
    - **Risk assessment feedback loops**

12.2.6 Challenges and Future Prospects

- Need for **multiplexed and high-throughput assays** for stacked GMOs
- Overcoming limitations in detecting **degraded DNA or protein** in processed foods
- Development of **portable and point-of-care diagnostics**
- Use of **artificial intelligence (AI)** for data interpretation and pattern recognition
- Incorporation of **CRISPR-based biosensors** and **lab-on-chip platforms** for real-time monitoring

12.2.7 Conclusion

Sampling and analytical techniques serve as the **bedrock of GMO biosafety and regulatory systems**. Scientific precision in sampling, coupled with robust, validated detection technologies, ensures the **traceability**, **accountability**, and **safety of transgenic applications**. As transgenic innovations continue to evolve, so must the detection landscape—with greater sensitivity, portability, and integration into **smart biosurveillance frameworks** that protect human and environmental health while promoting biotechnological progress.

12.3 Regulatory Labeling and Traceability Requirements

12.3.1 Introduction

With the increasing commercialization and global trade of genetically modified organisms (GMOs), the **regulatory labeling and traceability of transgenic products** have become critical components of biosafety governance. These frameworks ensure that consumers, regulators, and industries are informed of the presence of GMOs in food, feed, and other products. Labeling and traceability also facilitate **accountability**, **transparency**, and **compliance with national and international biosafety regulations**, particularly in the context of food safety, environmental protection, and ethical concerns.

12.3.2 Regulatory Objectives

The primary goals of labeling and traceability regulations are:

- To provide **informed choices** to consumers
- To ensure **compliance with biosafety and trade regulations**
- To enable **product recall**, if safety concerns arise
- To support **risk assessment and post-market monitoring**
- To maintain **public trust and transparency**

12.3.3 Labeling Requirements: National and International Perspectives

Labeling requirements vary significantly across countries, ranging from **mandatory disclosure** to **voluntary or threshold-based systems**.

A. India's Regulatory Framework

India has adopted a **mandatory labeling regime** for GM foods and products under the **Food Safety and Standards Authority of India (FSSAI)** and **Legal Metrology (Packaged Commodities) Rules**.

- As per the **FSSAI's Gazette Notification (2018)**, all food products containing genetically engineered ingredients must be labeled as **"Contains GMO/Ingredients derived from GMO"** if the GMO content exceeds **5%** by weight.
- The **Genetic Engineering Appraisal Committee (GEAC)** under the Ministry of Environment, Forest and Climate Change (MoEF&CC) oversees the release and monitoring of GMOs.
- The **Bureau of Indian Standards (BIS)** and **Customs Authorities** enforce GMO labeling for imports and exports.

B. Global Standards and Variability

1. **European Union (EU)**
   - **Strict mandatory labeling** if GMO content exceeds **0.9%**
   - Applies to **food, feed, and ingredients**, whether the GM content is detectable or not
   - Requires labeling of both **direct and indirect GM-derived products**
2. **United States**
   - The **National Bioengineered Food Disclosure Standard (NBFDS)** mandates **bioengineered food labeling**, but allows flexibility in formats (text, symbol, digital link)
   - Threshold set at **5%**, with exemptions for incidental additives
3. **Japan**
   - Requires labeling of **specific GM ingredients** exceeding **5%** content
   - Does not mandate labeling if DNA/protein is undetectable in the final product
4. **Brazil, Australia, China, and Others**
   - Maintain diverse thresholds (ranging from **1% to 3%**) for mandatory labeling
   - Differences reflect variations in **consumer policy, trade priorities, and risk perception**

12.3.4 Traceability Mechanisms

Traceability refers to the **systematic documentation and tracking** of GMO products through the production, processing, and distribution chain. It is essential for:

- Identifying the **origin and movement** of transgenic materials
- Enabling **efficient product recalls**
- Supporting **risk management and liability control**
- Facilitating **regulatory audits and compliance**

A. Key Components of Traceability Systems

1. **Unique Identification of GMO Events**
    - Each approved GMO is assigned an **event-specific identifier**
    - Enables differentiation between multiple transgenic lines or traits
2. **Segregation and Identity Preservation (IP)**
    - Maintenance of **GMO and non-GMO segregation** during cultivation, transport, and processing
    - Includes **physical barriers**, **documentation**, and **batch separation**
3. **Recordkeeping and Documentation**
    - Detailed logs of **seed source, field location, harvesting, transport, processing, and sale**
    - Essential for **audit trails** and **legal accountability**
4. **Chain-of-Custody Protocols**
    - Ensures that transgenic materials are tracked **from origin to endpoint**
    - Maintains data integrity across **supply chain stakeholders**

B. Technologies for Traceability

- **Barcoding and RFID tags** for tracking seed lots and product batches
- **Blockchain systems** for immutable, transparent, and distributed data storage
- **Database integration** with national biosafety clearinghouses and regulatory bodies

12.3.5 Challenges in Implementation

Despite the importance of labeling and traceability, several challenges persist:

- **Technical limitations** in detecting highly processed GM-derived materials

- **Lack of harmonization** across international standards complicates global trade
- **Costs of segregation and testing** for small-scale producers and exporters
- **Enforcement gaps** due to inadequate infrastructure or trained personnel
- **Consumer confusion** due to inconsistent or unclear labeling formats

12.3.6 Future Directions and Policy Recommendations

To enhance effectiveness and public trust in GMO labeling and traceability, future strategies may include:

- **Harmonization of international thresholds and labeling standards**
- **Public awareness campaigns** on GMO labeling implications
- **Strengthening laboratory and field testing infrastructure**
- **Encouraging adoption of digital traceability tools**
- **Capacity-building programs** for regulators, industry, and farmers

12.3.7 Conclusion

Labeling and traceability are fundamental tools in the **biosafety governance of transgenic organisms**. They uphold the principles of **consumer autonomy**, **risk management**, and **regulatory accountability**. While variability exists across countries, a concerted global effort toward **standardization, technological integration, and public engagement** will strengthen the transparency and acceptability of GMOs in agriculture, food systems, and international trade.

PART-IV

Bioethics and Societal Issues in Biotechnology

13

Field Trials and Standard Operating Procedures

13.1 Biosafety Research Trials: Scope and Implementation

Introduction

Biosafety research trials serve as a critical phase in the evaluation and regulation of genetically modified organisms (GMOs), particularly transgenic crops and microorganisms. These trials assess potential environmental and health impacts under confined and controlled conditions before the organism can be considered for commercial release. With the expansion of biotechnology applications, especially in agriculture, pharmaceuticals, and industrial processes, the need for scientifically rigorous and ethically sound biosafety research trials has become increasingly important.

13.1.1. Scope of Biosafety Research Trials

1.1. Definition and Objectives

Biosafety research trials are **confined, stepwise, and monitored field or laboratory-based studies** designed to evaluate the **safety, stability, performance, and unintended effects** of GMOs. The overarching objectives include:

- **Assessment of potential risks** to human health, non-target organisms, and the environment.
- **Evaluation of gene stability**, trait expression, and environmental adaptability.
- **Monitoring of gene flow** and potential invasiveness or weediness.
- **Determination of agronomic efficacy** under field conditions.
- **Data generation** for regulatory dossiers and risk assessment reports.

1.2. Key Domains

The scope of biosafety research trials encompasses various sectors:

- **Agricultural biotechnology**: GM crops (e.g., Bt cotton, GM mustard).

- **Medical biotechnology**: Transgenic animal models, recombinant vaccines.
- **Environmental biotechnology**: Genetically engineered microbes for bioremediation.
- **Industrial biotechnology**: Modified organisms for biofuel or enzyme production.

13.1.2. Types of Biosafety Research Trials

Biosafety trials are typically categorized by the level of containment and trial objective:

Type	Objective	Containment Level
Contained Laboratory Trials	Molecular characterization, gene expression	Physical containment (BSL-1 to BSL-3)
Confined Field Trials (CFTs)	Environmental interaction, agronomic traits	Biological & physical confinement
Multi-location Trials	Regional performance, ecological variability	Semi-confined
Large-scale Demonstration Trials	Near-commercial scale assessment	Regulatory oversight

13.1.3. Regulatory Framework in India

In India, the **Biosafety Research Trials** are governed by the **Rules, 1989 of the Environment Protection Act, 1986**, and are coordinated through several key authorities:

- **Institutional Biosafety Committees (IBSCs)** – Review research proposals within institutions.
- **Review Committee on Genetic Manipulation (RCGM)** – Grants approval for laboratory and CFTs.
- **Genetic Engineering Appraisal Committee (GEAC)** – Approves environmental release and large-scale trials.

All trials must comply with the **Biosafety Research Level-I (BRL-I)** and **Biosafety Research Level-II (BRL-II)** trial guidelines issued by the **Department of Biotechnology (DBT)**.

13.1.4. Implementation Process

The successful execution of biosafety research trials requires a multi-phased and well-regulated implementation framework:

4.1. Proposal Submission and Approval

- Submission of a detailed trial protocol to the **IBSC**, followed by RCGM review.

- Evaluation of construct stability, host range, gene expression, and trial design.
- Clearance for BRL-I trials after initial laboratory safety assessment.

4.2. Site Selection and Preparation

- Selection of isolated and secure locations to minimize gene flow.
- Establishment of buffer zones and monitoring plots.
- Use of border rows and reproductive containment (e.g., male sterility).

4.3. Trial Execution and Monitoring

- Strict compliance with **Standard Operating Procedures (SOPs)** and biosafety guidelines.
- Regular monitoring for phenotypic performance, gene expression, pest resistance, etc.
- Environmental sampling for unintended impacts.

4.4. Reporting and Data Submission

- Generation of **comprehensive biosafety and agronomic data**.
- Submission to RCGM/GEAC for risk assessment and regulatory review.
- Preparation of **confidential business information (CBI)** and public summaries for transparency.

13.1.5. Challenges in Implementation

Despite the robustness of guidelines, implementation often faces:

- **Lack of awareness** among field staff and local farmers.
- **Public mistrust** and opposition to GM trials.
- **Regulatory delays** and procedural bottlenecks.
- **Data management and traceability issues**.

13.1.6. Ethical and Societal Considerations

Conducting biosafety trials in a socially responsible manner involves:

- **Community engagement and prior informed consent**.
- **Transparency and public consultation**.
- **Ethical approval** from appropriate institutional committees.
- Ensuring **biosafety compliance** does not compromise **biodiversity or food sovereignty**.

13.1.7. Conclusion

Biosafety research trials form the cornerstone of **science-based decision-making** in biotechnology regulation. Their proper implementation ensures a **balance between innovation and safety**, safeguarding public health and environmental integrity. With evolving biotechnologies and rising societal concerns, these trials must continue to be **adaptive, transparent, and globally harmonized**.

13.2 Guidelines for Confined Field Trials (CFT)

Introduction

Confined Field Trials (CFTs) are **controlled outdoor evaluations** of genetically modified organisms (GMOs), particularly genetically modified (GM) crops, conducted to assess their **performance, safety, and environmental impact** before unrestricted release. CFTs play a crucial role in generating **empirical, science-based data** to support biosafety assessments. These trials are designed with stringent **biological and physical containment measures** to prevent gene flow and ensure environmental integrity.

In India and globally, CFTs are regulated through well-defined **biosafety guidelines and Standard Operating Procedures (SOPs)** under the framework of national and international biosafety protocols.

1. Objectives of CFT Guidelines

The primary objectives of the guidelines governing CFTs are to:

- **Ensure environmental and biological confinement** of transgenic material.
- **Prevent unintentional release** and gene flow into related species.
- **Facilitate scientific evaluation** of agronomic performance and biosafety.
- **Protect biodiversity**, human health, and socio-economic interests.
- Generate **statistically valid data** for regulatory decision-making.

2. Regulatory Authorities Involved

In India, CFTs are governed under the **Rules, 1989 of the Environment (Protection) Act, 1986**. The key regulatory authorities include:

- **Institutional Biosafety Committee (IBSC):** Initial trial review and compliance within the institution.
- **Review Committee on Genetic Manipulation (RCGM):** Approves BRL-I level CFTs and provides technical oversight.

- **Genetic Engineering Appraisal Committee (GEAC):** Responsible for granting permission for BRL-II trials and environmental release.
- **State Biotechnology Coordination Committees (SBCCs)** and **District Level Committees (DLCs):** Field-level monitoring and local compliance.

3. Key Elements of CFT Guidelines

The implementation of CFTs requires adherence to various structured components designed to maintain **biosafety integrity** throughout the trial period.

3.1. Site Selection Criteria

- Isolation from sexually compatible crops to prevent cross-pollination.
- Non-GMO buffer zones and border rows to mitigate pollen dispersal.
- Avoidance of ecologically sensitive, flood-prone, or protected areas.
- Secure access to prevent human or animal intrusion.

3.2. Application and Approval Process

- Submission of a detailed **CFT application dossier** to the RCGM/GEAC.
- Inclusion of molecular characterization, trait information, and containment strategy.
- Submission of a **signed agreement** ensuring compliance with biosafety and trial termination procedures.

3.3. Trial Design and Management

- Use of **replicated randomized block designs** for scientific rigor.
- Implementation of **biological confinement** (e.g., male sterility, temporal separation).
- Installation of **physical barriers** such as fencing or netting where necessary.
- Strict **documentation and labeling** of GM material.

3.4. Monitoring and Compliance

- Deployment of **DLCs and SBCCs** for periodic on-site inspections.
- Maintenance of detailed **field records, observations, and logs**.
- Immediate reporting of any unexpected outcomes (e.g., volunteer plants, pest susceptibility).

3.5. Post-Trial Management

- **Destruction of plant material** (harvested and residual) through incineration or deep burial.
- **Monitoring of the trial site** post-harvest for regrowth or volunteers for at least one season.
- No cultivation of the same or related species for a specified duration on the trial site.
- Submission of a **comprehensive trial report** to regulatory authorities.

4. Special Considerations

4.1. Socio-Environmental Sensitivity

- Conducting stakeholder consultations and engaging local communities when appropriate.
- Considering local biodiversity, indigenous varieties, and traditional knowledge systems.

4.2. Data Confidentiality and Public Disclosure

- Protection of Confidential Business Information (CBI) where warranted.
- Submission of non-confidential summaries for public access to promote transparency.

4.3. International Harmonization

The Indian guidelines for CFTs are aligned with global standards such as those from:

- **OECD Consensus Documents**
- **Codex Alimentarius Commission**
- **Cartagena Protocol on Biosafety**

These harmonizations facilitate **data portability** and **international acceptance** of Indian-generated biosafety data.

5. Challenges in Execution

- **Inadequate infrastructure** and trained personnel at local monitoring levels.
- **Public skepticism** and misinformation regarding GMOs.
- **Logistical challenges** in site isolation and containment, especially in fragmented landholdings.
- **Delay in approvals** leading to seasonal mismatches.

Conclusion

Confined Field Trials form a **scientific bridge** between laboratory research and commercial application of GMOs. Their stringent implementation ensures that **biosafety concerns are addressed proactively**, balancing innovation with precaution. Adherence to national CFT guidelines not only supports **evidence-based regulatory decision-making** but also bolsters public confidence in biotechnology governance.

13.3 Standard Operating Procedures (SOPs) for Risk Mitigation

Introduction

Standard Operating Procedures (SOPs) for risk mitigation are **predefined, validated, and regulatory-approved protocols** designed to systematically minimize potential risks associated with the field testing and handling of genetically modified organisms (GMOs). In the context of biosafety, particularly during **Confined Field Trials (CFTs)** and related experimental releases, SOPs are indispensable tools that guide responsible implementation, ensure compliance with biosafety regulations, and **protect human health, biodiversity, and the environment**.

Risk mitigation SOPs serve as operational blueprints that address both **anticipated and unanticipated hazards**, thereby embedding a culture of safety and accountability in all phases of GMO development and assessment.

1. Objectives of Risk Mitigation SOPs

- To **minimize the likelihood of gene flow** to non-target organisms or the environment.
- To ensure **safe containment, handling, transportation, and disposal** of GMOs.
- To establish **uniformity and consistency** in safety practices across research and trial sites.
- To facilitate **regulatory compliance** and documentation for biosafety audits.
- To support **traceability, transparency, and public trust** in transgenic research.

2. Key Components of SOPs for Risk Mitigation

SOPs are designed to encompass the full spectrum of activities involved in GMO field trials, from **pre-sowing to post-harvest** phases. The essential elements include:

2.1. Site Selection and Preparation

- Selection of isolated and secured trial plots with **adequate spatial and temporal distancing** from compatible crops.
- Preparation of **buffer zones and border rows** to serve as pollen traps.
- Erection of **physical barriers or fencing** to restrict unauthorized entry.

2.2. Personnel Training and Access Control

- Orientation and certification of all personnel on **biosafety guidelines, PPE usage, and trial-specific SOPs**.
- Controlled access to trial sites, maintained through **visitor logs and authorization systems**.
- Provision of **emergency response procedures** in case of protocol breaches or exposure incidents.

2.3. Planting, Maintenance, and Observation

- Sowing of GM crops using **identified and traceable seed lots**, with accurate documentation.
- Periodic monitoring for **phenotypic stability, pest interaction, and volunteer emergence**.
- Maintenance of **crop diaries** and trial-specific observation logs for each event.

2.4. Biological and Physical Containment Measures

- Enforcement of **biological confinement strategies**, including:
    - Male sterility,
    - Use of non-flowering plants,
    - Synchronized or asynchronous flowering regimes.
- Installation of **physical containment infrastructure** as required:
- Shade nets,
- Windbreaks,
- Guarded perimeters.

2.5. Post-Harvest Handling and Disposal

- Collection of harvestable material under **controlled and traceable conditions**.

- Prohibition of harvested GM material from entering **commercial, food, or feed chains**.
- Destruction of plant residues through **incineration, deep burial, or autoclaving**, as per biosafety norms.

2.6. Monitoring for Volunteers and Regrowth

- Continued surveillance of the trial site for a **minimum of one growing season** post-harvest.
- Manual removal and documentation of **volunteer plants** or regrowth from residual GM material.
- Restriction on planting the same or related species in the subsequent crop season on the same site.

2.7. Incident Response and Corrective Action

- Implementation of **pre-approved emergency protocols** in case of unintended release or exposure.
- Documentation of **non-compliance events**, followed by root cause analysis and remedial steps.
- Immediate reporting to **Institutional Biosafety Committees (IBSCs), State Boards, or GEAC**, as required.

2.8. Record-Keeping and Reporting

- Maintenance of detailed records including:
  - Source and quantity of GM seeds,
  - Trial maps and design,
  - Inspection reports,
  - Disposal logs.
- Submission of **periodic reports** to regulatory bodies in standardized formats, highlighting observations and compliance status.

3. Institutional Responsibilities in SOP Enforcement

SOPs are enforced and overseen by multiple layers of institutional and governmental frameworks:

- **Principal Investigators (PIs)** and Trial Managers are accountable for day-to-day compliance and implementation.
- **Institutional Biosafety Committees (IBSCs)** must verify SOPs before trial initiation and conduct interim assessments.

- **Monitoring Committees** (DLCs and SBCCs) ensure independent evaluation of SOP adherence at field sites.
- **Regulatory bodies** like RCGM and GEAC assess SOP execution as part of biosafety dossier reviews.

4. Harmonization with International Standards

The SOPs prescribed in India are aligned with global frameworks, including:

- **OECD Consensus Documents** on environmental safety,
- **ISAAA and AGBIOS guidelines** for risk assessment,
- **International Standards for Phytosanitary Measures (ISPMs)**.

This harmonization ensures **credibility and acceptability** of Indian biosafety data in international regulatory and trade contexts.

5. Challenges and Considerations

- **Variability in implementation** across regions due to resource constraints or lack of trained personnel.
- **Inadequate record maintenance** and deviation from SOPs under field conditions.
- Ensuring **stakeholder awareness and community engagement**, particularly in CFT zones.
- Need for **dynamic updates** of SOPs with advances in genetic engineering techniques and containment technologies.

Conclusion

Standard Operating Procedures (SOPs) for risk mitigation are the **cornerstones of biosafety assurance** in the field testing of GMOs. These guidelines not only uphold the **scientific integrity and reproducibility** of field trials but also reinforce **regulatory confidence, public transparency, and environmental stewardship**. A rigorous and ethically sound application of SOPs ensures that transgenic research progresses within a **responsible, risk-mitigated, and legally compliant** framework.

14

Bioethics in Biotechnology

14.1 Ethics in Scientific Research and Innovation

Introduction

Scientific research and innovation in biotechnology, particularly in the realms of genetic engineering, genomics, and transgenic organisms, bring immense promise for advancing agriculture, health, and environmental sustainability. However, this progress must be **balanced with strong ethical foundations**, as the misuse or careless application of biotechnological innovations can lead to **irreversible harm to humans, society, and the biosphere**. The term *bioethics* encapsulates the **principles of morality, justice, and responsibility** that should govern all scientific inquiry and technological development.

Ethics in scientific research and innovation is not merely an adjunct to regulatory compliance—it is a **core component of scientific integrity and social responsibility**. It ensures that biotechnological advances are developed in ways that respect human dignity, safeguard biodiversity, and reflect societal values.

1. Fundamental Ethical Principles in Scientific Research

Ethical decision-making in science is guided by several **foundational principles**, adapted from biomedical and research ethics. These include:

1.1 Respect for Persons

- Researchers must acknowledge the **autonomy and dignity** of individuals who may be affected by scientific innovation.
- Informed consent is mandatory for any human or community involvement in experimental trials.

1.2 Beneficence

- The goal of scientific research should be to **maximize benefits** and **minimize potential harm**.
- Risk-benefit analyses are critical in both experimental design and innovation deployment.

1.3 Non-Maleficence

- The principle of "do no harm" mandates **careful oversight** to prevent intentional or unintentional injury to people, animals, or ecosystems.

1.4 Justice

- Research must be **fair and equitable**, ensuring **equal access to benefits** and **no disproportionate risk** to vulnerable groups.
- This includes considerations of **intellectual property rights**, **indigenous knowledge**, and **biopiracy prevention**.

2. Ethical Challenges in Biotechnological Innovation

The rapidly evolving landscape of biotechnology has introduced **complex ethical dilemmas**, including:

2.1 Human Genetic Manipulation

- Genetic interventions in humans, especially **germline editing**, raise concerns about **designer babies, eugenics, and intergenerational impact**.
- Ethical consensus varies globally; most regulations currently **prohibit human germline modifications** for reproductive purposes.

2.2 Animal Welfare in Research

- Transgenic animal models are essential for biomedical research but must adhere to **strict ethical guidelines** to minimize pain and distress.
- The **3Rs Principle (Replacement, Reduction, Refinement)** governs ethical animal experimentation.

2.3 Environmental Ethics

- Introducing genetically modified organisms (GMOs) into ecosystems may lead to **irreversible ecological disturbances**.
- Ethical stewardship demands **precautionary approaches, biodiversity conservation**, and transparent environmental impact assessments.

2.4 Dual-Use Dilemma

- Biotechnology tools, like CRISPR or synthetic biology, have potential for **both beneficial and harmful applications** (e.g., bioweapons).
- Scientists and institutions must evaluate and restrict **dual-use research of concern (DURC)**.

2.5 Data Privacy and Genetic Information

- With increasing use of genomic data, there is a pressing need to protect **individual privacy and data ownership**, especially in **population genomics and personalized medicine**.

3. Institutional Mechanisms to Uphold Ethics

To embed ethical values into biotechnology research, several institutional frameworks are established at national and global levels:

3.1 Institutional Ethics Committees (IECs)

- Every research institution must have an **ethics committee** to scrutinize project proposals involving **humans, animals, or GMOs**.
- These bodies review the **scientific rationale, risk assessment, and ethical implications** before granting approval.

3.2 Codes of Scientific Conduct

- Guidelines issued by organizations such as **ICMR, DBT, CSIR, and UNESCO** promote **ethical research behavior**, including:
  - Honesty in data reporting,
  - Avoidance of plagiarism,
  - Responsible authorship,
  - Proper data management.

3.3 Ethics in Funding and Publication

- Funding agencies increasingly require **ethical clearance certificates**.
- Peer-reviewed journals enforce **disclosure of conflicts of interest**, ethical declarations, and adherence to reporting guidelines (e.g., ARRIVE, CONSORT).

4. The Role of Education and Ethical Literacy

Creating a culture of ethical science necessitates that students and researchers are **educated in ethical reasoning and moral responsibility**:

- Courses in **bioethics, IPR, and regulatory science** are now integral to postgraduate biotechnology curricula.
- Institutions conduct **workshops, seminars, and training** to promote awareness of emerging ethical issues.

5. Global Perspectives on Bioethics

Many **international treaties and conventions** help shape ethical biotechnology, such as:

- **Universal Declaration on Bioethics and Human Rights (UNESCO, 2005)**,
- **Convention on Biological Diversity (CBD)** and **Cartagena Protocol**,
- **World Medical Association's Helsinki Declaration**,
- **Nagoya Protocol on Access and Benefit Sharing (ABS)**.

These emphasize that **science must be guided by human rights, equity, sustainability**, and public accountability.

Conclusion

Ethics in scientific research and innovation is not merely a regulatory necessity but a **moral imperative** that safeguards society, future generations, and the planet. In the context of biotechnology, ethics must be seen as **a compass guiding responsible discovery**, ensuring that innovation aligns with **public good, justice, and environmental sustainability**. Only through rigorous ethical scrutiny, transparency, and accountability can biotechnology truly serve humanity in a fair and sustainable manner.

14.2 Religious, Social, and Cultural Perspectives

Introduction

The intersection of biotechnology with society transcends science and regulation—it deeply engages with **religious, social, and cultural values** that shape public acceptance, ethical legitimacy, and policy decisions. While scientific advancement often follows the logic of innovation and utility, **religious and cultural frameworks** evaluate it through lenses of morality, sanctity of life, spiritual doctrine, and collective well-being. In multicultural societies like India and across the globe, these perspectives influence how genetically modified organisms (GMOs), cloning, stem cell research, and transgenic interventions are perceived and adopted.

Understanding and respecting **diverse worldviews** is not merely a matter of ethical courtesy; it is crucial for **social sustainability, participatory governance**, and long-term acceptance of biotechnological innovations.

1. Religious Perspectives on Biotechnology

1.1 Hinduism

- Hindu philosophy emphasizes **reverence for life (ahimsa), the interconnectedness of all beings (vasudhaivakutumbakam)**, and balance with nature.
- Genetic modification of plants may be acceptable when used for the **welfare of society (lokasangraha)**, but there is ethical unease over:

- Use of animal genes in plants or humans,
- Cloning and manipulation of embryos,
- Perceived disrespect toward natural order (*prakriti*).

1.2 Islam

- Islamic bioethics is rooted in the **Qur'an and Shariah law**, with central principles of **halal (permissibility), haram (prohibition), and maslahah (public interest)**.
- Genetic modifications are acceptable if they serve health and public good, do not alter human essence, and avoid prohibited sources (e.g., pig genes).
- **Consent, precaution, and the principle of "no harm" (la dararwa la dirar)** guide permissible practices.

1.3 Christianity

- Christian ethical concerns arise from doctrines of **natural law, divine creation**, and the sanctity of life.
- While many denominations support therapeutic use of biotechnology (e.g., for disease resistance or hunger alleviation), strong opposition exists on:
  - Germline editing,
  - Human cloning,
  - Patent ownership over life forms.

1.4 Buddhism

- Buddhism emphasizes **non-violence (ahimsa), compassion, and karma**.
- The use of biotechnology is evaluated by its **intent and consequence**—if it promotes well-being without causing suffering, it may be acceptable.
- There is aversion to harming sentient life forms and disrupting ecological balance.

1.5 Jainism

- Strict adherence to **non-violence and purity** guides Jain ethics.
- Any genetic manipulation that potentially harms microorganisms, animals, or disturbs ecological systems is seen as ethically problematic.

Conclusion from Religious Perspectives: There is **conditional acceptance** of biotechnology across religions, particularly when aligned with public welfare, environmental protection, and spiritual principles. However, **moral limits** are emphasized in areas involving cross-species genetic transfer, embryonic manipulation, and commodification of life.

2. Social Perspectives on Biotechnology

2.1 Public Perception and Awareness

- The **general public's understanding of biotechnology** is often limited, leading to fear, skepticism, or over-enthusiasm.
- Misinformation, especially regarding GM foods or vaccines, has led to **social resistance or activism**.
- Engaging communities through **science communication, public consultations, and participatory research** helps bridge the gap.

2.2 Equity and Access

- Biotechnology can exacerbate **social inequalities** if access to its benefits (e.g., high-yield seeds, diagnostics) is limited to affluent groups or multinational corporations.
- **Marginalized farmers, tribal communities**, and resource-poor populations may be excluded from decision-making or fair sharing of biotechnological benefits.

2.3 Gender Dimensions

- Women, especially in rural and agrarian societies, are deeply involved in **seed selection, food preparation, and biodiversity conservation**.
- Biotechnological decisions that ignore women's knowledge systems and roles can lead to **ethical oversights and disempowerment**.

3. Cultural Perspectives and Indigenous Knowledge

3.1 Cultural Sensitivities in Biotechnology

- Cultural norms and taboos often influence the **acceptability of certain biotechnologies**, such as:
  - Use of animal-derived products in vegetarian societies,
  - Interference with sacred species or ritual crops,
  - Introduction of foreign genes that violate cultural purity beliefs.

3.2 Role of Traditional Knowledge Systems

- Indigenous knowledge systems (IKS) offer **valuable ecological insights**, crop varieties, and conservation practices.

- Ethical biotechnology must ensure **protection of IKS, recognition of community rights, and benefit sharing** under frameworks like:
  - **Convention on Biological Diversity (CBD),**
  - **Nagoya Protocol on Access and Benefit Sharing.**

3.3 Biopiracy and Cultural Appropriation

- Unsanctioned use of indigenous genetic resources or traditional medicinal knowledge by companies or researchers leads to **biopiracy**.
- Ethical concerns include **loss of sovereignty, cultural exploitation, and unjust commercial gains**.

4. Integration into Ethical Governance

- Ethical governance in biotechnology requires **multi-stakeholder dialogue**, including **religious leaders, social activists, indigenous communities, and cultural experts**.
- Policymaking should be **inclusive, transparent, and culturally responsive**, acknowledging **pluralistic values** while maintaining scientific integrity.
- Initiatives like **community biosafety education, local ethics review boards**, and **ethnographic impact assessments** are encouraged.

Conclusion

Religious, social, and cultural perspectives represent the **moral fabric of societies** and serve as essential checks and balances on scientific enterprise. Ethical biotechnology must not operate in isolation from these perspectives but should be enriched by **dialogue, mutual respect, and cultural sensitivity**. Bridging the gap between laboratory innovation and societal values will foster **public trust, ethical resilience**, and **responsible science for global good**.

14.3 Environmental Ethics and Biodiversity Protection

Introduction

Environmental ethics refers to the **moral principles governing the relationship between human beings and the natural environment**, emphasizing our duty to preserve ecological balance and protect biodiversity. In the context of biotechnology, particularly genetic engineering and transgenic interventions, environmental ethics serves as a **crucial evaluative framework** for ensuring that scientific progress does not compromise the health, diversity, or integrity of ecosystems.

The rise of modern biotechnologies has introduced **unprecedented opportunities**—from pest-resistant crops to synthetic biology—but also raised concerns about **unintended ecological consequences**, **loss of biodiversity**, and **ethical responsibility** toward future generations. Thus, environmental ethics demands that biotechnology be guided not only by utility but also by **stewardship, sustainability, and respect for life.**

1. Ethical Foundations in Environmental Protection

Environmental ethics is grounded in several key philosophical paradigms:

- **Anthropocentric View**: Focuses on human benefits; nature is valued for its utility. Biotechnology is ethical if it serves human welfare without excessive environmental damage.
- **Ecocentric View**: Places intrinsic value on all life forms and ecosystems. Ethical biotechnology must ensure **minimal disturbance to ecological harmony**.
- **Biocentric View**: Recognizes the moral standing of all living beings. Interventions such as gene editing or release of GMOs must respect the **sanctity of non-human life**.
- **Deep Ecology**: Advocates for a radical shift from exploitation to reverence for nature. Biotechnology is acceptable only when it supports **ecological balance and interdependence**.

These frameworks underscore the need for **precautionary and responsible biotechnological practices**.

2. Biotechnology and Biodiversity: Risks and Responsibilities

Biodiversity is the **foundation of ecological resilience**, providing genetic resources for food, medicine, and environmental stability. However, **unregulated or poorly assessed biotechnological interventions** may pose significant threats:

2.1 Genetic Uniformity and Loss of Native Varieties

- Transgenic crops may displace **traditional cultivars**, leading to **genetic erosion**.
- Commercial monocultures reduce agro-biodiversity, making ecosystems **more vulnerable to pests, climate stress, and disease**.

2.2 Unintentional Spread and Gene Flow

- Genes from GMOs may **introgress into wild relatives** through pollen dispersal.
- This could result in **weedy or invasive hybrids**, threatening native flora and altering community structures.

2.3 Non-target Impacts and Ecological Disruption

- GM crops producing insecticidal proteins (e.g., Bt toxins) may harm **beneficial insects, soil microbiota, or aquatic organisms**.
- Long-term ecological studies are essential to **evaluate cascading impacts on food webs and habitats**.

2.4 Ethical Imperatives for Conservation

- Biotechnological interventions must uphold **international biodiversity protection protocols**, such as:
  - The **Convention on Biological Diversity (CBD)**,
  - The **Cartagena Protocol on Biosafety**,
  - The **Nagoya Protocol on Access and Benefit Sharing**.
- Conservation strategies should integrate **in situ and ex situ approaches**, respecting **local ecological knowledge and genetic diversity**.

3. The Precautionary Principle and Environmental Stewardship

The **Precautionary Principle** is central to environmental ethics in biotechnology. It advocates that:

"When an activity raises threats of harm to the environment or human health, precautionary measures should be taken even if some cause and effect relationships are not fully established scientifically."

In practice:

- Risk assessments must precede any release of GMOs into the environment.
- **Post-release monitoring**, impact modeling, and **adaptive management** are required to ensure environmental safety.
- Ethical biotech development must adopt a **"do no harm" philosophy**, combining innovation with **ecological mindfulness**.

4. Biodiversity Protection through Ethical Biotechnology

Contrary to causing harm, biotechnology can also support biodiversity if **appropriately directed**:

4.1 Conservation Genomics

- Molecular tools assist in **identifying and conserving endangered species**, characterizing gene pools, and preventing inbreeding.

4.2 Bioremediation

- Engineered microbes or plants (e.g., transgenic *Arabidopsis* or *Pseudomonas*) help detoxify polluted environments, restoring **degraded habitats**.

4.3 Sustainable Agriculture

- Genetic improvements can **reduce pesticide use**, improve stress tolerance, and **conserve soil and water resources**—thereby supporting biodiversity.
- **Participatory breeding** involving indigenous farmers helps preserve **landrace diversity** and traditional ecological practices.

4.4 Ex Situ Genetic Resource Banks

- Ethical biotech ensures the **long-term conservation of genetic materials** through cryopreservation, seed banks, and cell culture repositories.

Note: All such applications must be **transparent, regulated, and inclusive of local stakeholders**, ensuring **ethical sourcing, benefit sharing**, and **respect for indigenous rights**.

5. Ethical Governance and Global Environmental Justice

Environmental ethics in biotechnology must transcend borders and reflect the principles of **global environmental justice**:

- **Equitable access** to biodiversity-derived technologies and benefits must be guaranteed.
- **Developing countries** should not become testing grounds for unapproved transgenic technologies.
- **Community rights**, especially of tribal and marginalized populations, must be **legally protected** from bio-exploitation or ecological degradation.

Collaborative governance frameworks involving **scientists, policymakers, ethicists, and community representatives** are vital for ethically sound biotechnological innovation.

Conclusion

Environmental ethics and biodiversity protection form the **moral backbone** of responsible biotechnology. In the face of accelerating technological advancement, **ethical vigilance, ecological awareness, and biodiversity stewardship** must remain paramount. A balanced approach—rooted in sustainability, fairness, and scientific humility—will ensure that biotechnology serves both **humanity and the biosphere** in enduring harmony.

14.4 Ethical Guidelines for Genetic Engineering and GMOs

Introduction

Genetic engineering, especially in the context of **genetically modified organisms (GMOs)**, has significantly advanced modern science and agriculture. However, this technological progress brings with it **complex ethical dilemmas** that call for a robust set of guidelines to balance **innovation with responsibility, safety with freedom, and benefits with risks**. Ethical guidelines for genetic engineering and GMOs are designed to **safeguard human health, protect the environment, ensure social justice**, and uphold the **dignity of life** in all forms.

These guidelines form a **normative framework** that integrates **bioethical principles, legal standards, scientific assessments, and societal values**. Their implementation ensures that the development and deployment of genetic technologies are **transparent, inclusive, and ethically defensible**.

1. Core Ethical Principles in Genetic Engineering

The following fundamental bioethical principles guide all genetic engineering practices:

- **Autonomy**: Respect for individual and community rights, especially in cases where GMOs are introduced into food systems or ecosystems.
- **Beneficence**: Biotechnology must aim to do good—improving health, food security, or environmental quality—while maximizing benefits.
- **Non-Maleficence**: The principle of "do no harm" mandates that the risks of genetic interventions must be minimized or avoided altogether.
- **Justice**: Equitable access to benefits and fair distribution of risks, ensuring that no community, region, or group bears disproportionate burdens.
- **Precautionary Principle**: In the face of scientific uncertainty or potential irreversible harm, precaution must guide action.
- **Stewardship**: Humans have an ethical responsibility to protect natural ecosystems and biodiversity.

2. Ethical Frameworks in GMO Development

Ethical oversight must be integrated across **all stages of GMO development**, from laboratory research to commercialization. This includes:

2.1 Research and Development

- Prior to initiating genetic engineering research, **ethical review boards** must assess the objectives, potential impacts, and necessity of the intervention.

- Research involving **transgenic animals, plants, or microbes** must follow **institutional biosafety and ethical protocols**.
- Intellectual honesty and transparency are essential, especially in **publishing data** related to biosafety, gene stability, and efficacy.

2.2 Field Testing and Trials

- Trials must ensure **informed consent from affected communities**, and adhere to **confined field trial guidelines**.
- Ethical protocols must address issues like **environmental containment, gene flow prevention**, and **impact on local biodiversity**.

2.3 Commercialization and Deployment

- GMOs must not be released until **regulatory approvals, environmental impact assessments**, and **public consultations** are satisfactorily conducted.
- **Labeling and traceability** are ethical obligations to ensure **consumer autonomy** and informed choice.
- Special attention must be paid to **socio-economic consequences**, such as impacts on smallholder farmers, market access, and seed sovereignty.

3. Ethical Guidelines by National and International Bodies

Several national and global institutions have framed **ethical and biosafety guidelines** for GMOs. These include:

3.1 National Guidelines (India)

- **Department of Biotechnology (DBT)** and **Genetic Engineering Appraisal Committee (GEAC)** under the **Ministry of Environment, Forest and Climate Change** regulate GMO approvals and safety protocols.
- **Institutional Biosafety Committees (IBSCs)** oversee research at the institutional level.
- India mandates **environmental risk assessments**, **socio-economic impact studies**, and **public hearings** before GMO release.

3.2 International Protocols

- **Cartagena Protocol on Biosafety (2000)**: Ensures safe handling, transport, and use of living modified organisms (LMOs) with a focus on **transboundary movement**.
- **Nagoya Protocol (2010)**: Upholds **fair and equitable benefit-sharing** of genetic resources and associated traditional knowledge.

- **FAO/WHO Codex Alimentarius**: Provides ethical and safety standards for **GM food assessment**, including **allergenicity and toxicity testing**.

4. Key Ethical Concerns in Genetic Engineering

4.1 Unintended Consequences

- Genetic interventions may lead to **off-target effects**, **gene silencing**, or **horizontal gene transfer**, raising safety concerns.
- Ethical guidelines insist on **rigorous scientific validation** and **long-term ecological monitoring**.

4.2 Impact on Farmers and Indigenous Communities

- The introduction of GMOs may disrupt traditional farming systems or impose **intellectual property restrictions**.
- Ethical norms demand **prior informed consent (PIC)** and protection of **farmers' rights and seed sovereignty**.

4.3 Patenting and Ownership

- Ethical issues arise from **patenting of genes, transgenic organisms, or biotech tools**, often excluding indigenous knowledge holders.
- Guidelines support **open-access research**, **bioprospecting transparency**, and **equitable benefit sharing**.

4.4 Biosecurity and Dual-Use Risks

- Genetic technologies may be misused for bioterrorism or unethical applications.
- Ethical governance must include **biosecurity training**, **risk mitigation protocols**, and **international collaboration** to prevent misuse.

5. Ethical Review and Public Accountability

- Every GMO project must be subject to **ethical scrutiny by Institutional Ethics Committees (IECs)** and **biosafety authorities**.
- There must be **public access to information**, stakeholder engagement, and avenues for **grievance redressal**.
- Ethical guidelines emphasize **transdisciplinary collaboration** between scientists, ethicists, legal experts, and civil society to ensure **inclusive and responsible innovation**.

Conclusion

Ethical guidelines for genetic engineering and GMOs are indispensable to ensure that biotechnology progresses **in harmony with human values, ecological**

integrity, and social justice. These guidelines **act as moral compasses**, navigating the complexities of transgenic science with responsibility, care, and vision. As biotechnology continues to shape the future, its governance must remain **ethically grounded, socially responsive, and environmentally mindful**, affirming a shared commitment to sustainability and the common good.

15

Biopiracy and Ethical Misappropriation

15.1 Biopiracy – Definition, Examples, and Impacts

Introduction

Biopiracy represents a grave ethical and legal concern in the field of biotechnology and intellectual property rights. It refers to the **unauthorized and uncompensated exploitation of biological resources and traditional knowledge**—primarily from developing countries—by corporations, research institutions, or individuals, often for commercial gains. Biopiracy is inherently unjust as it involves the **misappropriation of community knowledge**, undermining the rights of **indigenous peoples**, and **violating the principles of equity, fairness, and sovereignty** over natural resources.

As globalization intensifies biotechnological exploration, biopiracy challenges ethical norms and highlights the need for **strong legal, institutional, and ethical safeguards**.

Definition of Biopiracy

Biopiracy can be defined as:

"The **unauthorized extraction, use, and commercialization of genetic resources or associated traditional knowledge** from indigenous communities or biodiversity-rich regions without **prior informed consent** or **fair and equitable benefit sharing**."

Biopiracy often involves patenting biological materials or indigenous medicinal knowledge without recognizing the **origin communities as rightful holders** of such knowledge or genetic wealth. It is closely linked to ethical misappropriation, exploitation, and **violations of bioethical norms and international agreements**.

Notable Examples of Biopiracy

1. The Neem Patent (Azadirachtaindica) – India

- In the 1990s, the **European Patent Office (EPO)** granted a patent to **W.R. Grace** and the **US Department of Agriculture** for an antifungal formulation derived from neem seeds.

- Indian scientists and environmentalists challenged the patent, arguing that neem has been used for centuries in Indian traditional medicine.
- In 2000, the EPO **revoked the patent**, recognizing the existence of **prior traditional knowledge**.

2. Turmeric Patent (Curcuma longa) – India

- The **University of Mississippi Medical Center** received a US patent for turmeric's wound-healing properties.
- The **Council of Scientific and Industrial Research (CSIR), India**, challenged the patent on the basis of existing Ayurvedic literature.
- The patent was **revoked** in 1997, as the use of turmeric was **traditional and well documented**.

3. Basmati Rice Patent – India

- The US company **RiceTec Inc.** was granted a patent for "Basmati" rice lines and grains.
- The patent claimed novelty based on aroma, grain length, and cooking properties.
- The Indian government contested the claim, arguing that **Basmati is a traditional crop** grown in the Indo-Gangetic plains.
- Following international pressure, several claims were withdrawn, though **geographical indication (GI) protection** was also advocated.

4. Ayahuasca Plant – Amazon Tribes

- A US citizen was granted a patent for the Ayahuasca plant (*Banisteriopsiscaapi*), used in sacred rituals by Amazonian tribes.
- Indigenous groups protested, arguing this amounted to **cultural and spiritual theft**.
- The patent was ultimately **revoked**, highlighting concerns over **cultural biopiracy**.

Impacts of Biopiracy

1. Ethical and Cultural Erosion

- Biopiracy**undermines the spiritual, cultural, and intellectual heritage** of indigenous and local communities.
- It ignores the **collective custodianship** of traditional knowledge and violates **communal rights**.

2. Economic Exploitation

- Companies and researchers profit from biopirated resources without **compensating source communities**, leading to **economic injustice**.
- Lack of benefit sharing deprives these communities of **livelihood opportunities and royalties**.

3. Loss of Sovereignty

- Biopiracy threatens **national and community sovereignty over biodiversity**.
- Developing nations often lose control over **native genetic resources**, impacting conservation and development.

4. Scientific Misconduct

- When research is conducted without **prior informed consent (PIC)** and **ethical clearance**, it constitutes **violation of ethical standards** and **scientific integrity**.
- It damages the credibility of global research and undermines **mutual trust** between communities and scientific institutions.

5. Hindrance to Conservation and Sustainable Use

- Biopiracy discourages indigenous communities from **sharing knowledge or preserving biodiversity**, fearing exploitation.
- It impedes the objectives of the **Convention on Biological Diversity (CBD)**, which emphasizes **access and benefit sharing (ABS)**.

Legal and Institutional Response

In response to increasing incidents of biopiracy, various national and international frameworks have been developed:

- **Convention on Biological Diversity (CBD), 1992**: Recognizes **sovereign rights of nations over their genetic resources**, and mandates **prior informed consent and fair benefit sharing**.
- **Nagoya Protocol on Access and Benefit Sharing, 2010**: Strengthens mechanisms for **equitable sharing of benefits** arising from the use of genetic resources and traditional knowledge.
- **Traditional Knowledge Digital Library (TKDL), India**: An initiative by CSIR and Ministry of AYUSH to **digitally document traditional knowledge** to prevent **wrongful patent claims**.
- **Protection of Plant Varieties and Farmers' Rights Act (PPV&FRA), 2001 – India**: Protects the rights of traditional farmers and ensures **benefit sharing from plant varieties**.

- **Geographical Indication (GI) Tagging**: Provides recognition and protection to region-specific biological products.

Conclusion

Biopiracy is not just a legal issue, but a **deeply ethical concern** that challenges the very foundation of **equity, respect, and justice** in scientific research and international commerce. Addressing biopiracy requires a **multifaceted strategy** involving **ethical awareness, community empowerment, legal enforcement, and international cooperation**. Postgraduate students and future biotechnologists must recognize that **scientific innovation must not come at the cost of exploitation**, but should promote **inclusive, transparent, and responsible biotechnology** for the benefit of all.

15.2 International Conventions Against Biopiracy (CBD, Nagoya Protocol)

Introduction

The growing concern over biopiracy—defined as the unauthorized use and commercialization of biological resources and associated traditional knowledge—has led to the formulation of robust international legal instruments. These conventions are primarily aimed at ensuring **sovereign rights over biological resources**, promoting **fair and equitable benefit sharing**, and **protecting indigenous and local communities** from exploitation.

Among the most significant global frameworks combating biopiracy are the **Convention on Biological Diversity (CBD)** and the **Nagoya Protocol on Access and Benefit Sharing (ABS)**. These agreements emphasize ethical research, legal access mechanisms, and equitable collaboration in the use of biodiversity.

1. The Convention on Biological Diversity (CBD), 1992

Overview

The **Convention on Biological Diversity (CBD)** was adopted at the Earth Summit in Rio de Janeiro in 1992. It is the first comprehensive international agreement addressing **conservation of biodiversity, sustainable use of its components,** and **fair and equitable sharing of benefits** arising from the utilization of genetic resources.

Objectives

The CBD rests on three fundamental pillars:

- **Conservation** of biological diversity.
- **Sustainable use** of its components.

- **Fair and equitable sharing** of benefits arising from the use of genetic resources.

Key Provisions Relevant to Biopiracy

1. **National Sovereignty Over Resources (Article 3):**
   - Countries have sovereign rights over their biological resources and are responsible for ensuring conservation and sustainable use.
2. **Access to Genetic Resources (Article 15):**
   - Access to genetic resources shall be subject to **prior informed consent (PIC)** of the country providing the resource.
   - Access must be based on **mutually agreed terms (MAT)** to ensure fair benefit sharing.
3. **Recognition of Traditional Knowledge (Article 8(j)):**
   - Encourages respect, preservation, and maintenance of **indigenous knowledge**.
   - Promotes the involvement of indigenous communities in biodiversity conservation and benefit sharing.
4. **Technology Transfer and Capacity Building (Articles 16–19):**
   - Developed countries are encouraged to facilitate access to technologies and build capacities in biodiversity-rich countries.

CBD's Role in Combating Biopiracy

- By recognizing national sovereignty and enforcing **PIC and ABS mechanisms**, the CBD offers legal recourse against unauthorized exploitation.
- It has prompted many countries to adopt **national biodiversity laws**, ethical codes, and institutional frameworks.

2. The Nagoya Protocol on Access and Benefit Sharing, 2010

Overview

Adopted in 2010 at the 10th Conference of the Parties (COP-10) to the CBD in Nagoya, Japan, the **Nagoya Protocol** is a legally binding supplementary agreement to the CBD. It provides a **transparent legal framework** for the **effective implementation of access and benefit-sharing** provisions.

Objectives

- To ensure **fair and equitable sharing of benefits** arising from the utilization of genetic resources.
- To strengthen the **legal certainty, transparency, and compliance mechanisms** related to ABS.

Core Principles

1. **Prior Informed Consent (PIC):**
   - Genetic resources and associated traditional knowledge may only be accessed with the **explicit, informed permission** of the resource-providing country and/or community.
2. **Mutually Agreed Terms (MAT):**
   - Contracts or agreements must be established between resource providers and users, stipulating **terms for benefit sharing**, commercial use, intellectual property rights, and dispute resolution.
3. **Benefit Sharing:**
   - Benefits may be **monetary (royalties, licensing fees)** or **non-monetary (technology transfer, joint research, training)**.
   - Benefits should support conservation, community development, and traditional knowledge systems.
4. **Compliance Measures:**
   - Countries must designate **National Focal Points (NFPs)** and **Competent National Authorities (CNAs)**.
   - Users of genetic resources must provide **evidence of legal access** and **respect ABS obligations**.

Implementation Mechanisms

- **ABS Clearing-House Mechanism:** Facilitates exchange of information related to national ABS policies, permits, and agreements.
- **Monitoring and Enforcement:** Parties must establish checkpoints and penalties for non-compliance, thus preventing biopiracy.

Impact of CBD and Nagoya Protocol on Global Biopiracy Prevention

Dimension	Impact
Legal Empowerment	Ensures countries have **sovereign rights** to regulate access to their biodiversity.
Ethical Research	Mandates ethical standards like **PIC**, **MAT**, and **community participation**.
Recognition of Indigenous Rights	Validates **customary laws and knowledge systems** of indigenous communities.
Institutional Development	Encourages the establishment of **national ABS frameworks**, databases (e.g., TKDL), and monitoring authorities.
International Collaboration	Promotes fair partnerships between biotechnological corporations and resource-supplying countries.

Challenges in Implementation

Despite their strengths, CBD and the Nagoya Protocol face several practical challenges:

- **Lack of awareness and capacity** in biodiversity-rich countries.
- **Ambiguities in defining traditional knowledge and ownership.**
- **Weak enforcement mechanisms** at the national level.
- **Tensions between open-access scientific research and ABS compliance.**
- **Delayed ratification and non-participation** by some countries (e.g., USA is not a party to CBD).

Conclusion

The **CBD and the Nagoya Protocol** together form the backbone of the global legal and ethical response to biopiracy. They uphold **sovereignty, equity, and justice** in the utilization of genetic resources and traditional knowledge. However, effective prevention of biopiracy depends on **robust national implementation**, **empowerment of local communities**, **scientific integrity**, and **international cooperation**.

For postgraduate students in biotechnology and IPR, understanding these conventions is crucial not only for legal compliance but also to uphold the ethical values that must underpin modern biotechnology research and commercialization.

15.3 Ethical Approaches to Benefit Sharing and Indigenous Knowledge Protection

1. Introduction

The ethical imperative to protect indigenous knowledge (IK) and ensure fair and equitable benefit sharing is central to sustainable and respectful biotechnology practices. Traditional knowledge, especially related to biodiversity, agriculture, and medicinal use of plants, has been **accumulated over generations** by indigenous and local communities. This knowledge is not only invaluable for scientific discovery but is also a vital component of **cultural heritage and identity**.

However, unethical appropriation—often termed **biopiracy**—has raised serious concerns regarding justice, consent, and equity. Thus, ethical frameworks and legal instruments have evolved to safeguard **community rights**, promote **equitable benefit sharing**, and recognize the **intellectual contributions** of indigenous peoples.

2. Foundations of Ethical Benefit Sharing

Ethical benefit sharing refers to a system wherein benefits—monetary or non-monetary—arising from the utilization of genetic resources and associated knowledge are **equitably distributed** among all stakeholders, particularly the knowledge holders.

Core Ethical Principles:

- **Justice and Equity:** Ensuring that benefits do not accrue solely to corporations or researchers, but are equitably shared with resource providers and communities.
- **Prior Informed Consent (PIC):** No access to traditional knowledge or genetic resources should be permitted without **voluntary, prior, and informed consent** of the community.
- **Mutually Agreed Terms (MAT):** Agreements must be established outlining the **conditions of access**, **use**, and **benefit sharing** in a transparent and mutually respectful manner.
- **Respect for Cultural Integrity:** Protection of IK should include **preservation of traditions, beliefs, and customary laws** that govern knowledge transmission.
- **Empowerment and Participation:** Ethical systems must encourage the **active involvement of indigenous communities** in decision-making and policy formulation.

3. Forms of Benefit Sharing

Benefit sharing can be structured in multiple forms, both **monetary** and **non-monetary**, ensuring ethical reciprocity:

A. Monetary Benefits:

- Royalties from commercialization of products derived from IK.
- Licensing fees and access fees.
- Equity in companies or product lines developed from indigenous knowledge.
- Direct payments to community trust funds.

B. Non-Monetary Benefits:

- Capacity building and training programs.
- Joint research ventures with local institutions.
- Infrastructure development in indigenous communities (e.g., schools, clinics).

- Technology transfer and improved access to scientific tools.
- Recognition through authorship, patents, and co-inventorship where applicable.

4. Ethical Protection of Indigenous Knowledge

The ethical protection of IK requires **a shift from extractive research** models to **participatory, respectful, and community-centered approaches**. This includes:

A. Recognition of Collective Ownership:

- Indigenous knowledge is **communal**, not individual, and ethical frameworks must respect this collective identity.
- Legal systems should evolve to recognize **customary rights and knowledge systems** that fall outside conventional IPR definitions.

B. Documentation and Defensive Protection:

- Community-led documentation of traditional knowledge helps in **preventing misappropriation** and supports evidence in legal disputes.
- Databases like India's **Traditional Knowledge Digital Library (TKDL)** serve as powerful tools to prevent patenting of pre-existing knowledge.

C. Community Protocols and Registers:

- Ethical frameworks encourage the development of **Community Biodiversity Registers (CBRs)** and **Biocultural Community Protocols (BCPs)** which outline the terms for access and use of their knowledge.

D. Integration with Customary Laws:

- Ethical protection of IK must respect **customary rules, rituals, and community sanctions** that regulate knowledge access.
- Such an approach bridges the gap between **state legislation** and **community governance**.

5. Role of International Instruments in Ethical Approaches

Several global frameworks emphasize ethical principles in benefit sharing and knowledge protection:

- **Nagoya Protocol (2010):** Strongly promotes equitable benefit sharing and recognizes community rights to grant PIC and negotiate MAT.
- **UN Declaration on the Rights of Indigenous Peoples (UNDRIP, 2007):** Affirms the rights of indigenous peoples to maintain, control, and protect their cultural heritage and traditional knowledge.

- **World Intellectual Property Organization (WIPO):** Engaged in the development of an **International Legal Instrument on Intellectual Property and Genetic Resources, Traditional Knowledge and Folklore**.

6. Case Studies: Ethical vs Unethical Practices

Case	Nature	Ethical Evaluation
Kani Tribe and Jeevani (India)	Traditional knowledge on *Arogyapacha* used to develop Jeevani medicine.	**Ethical**, benefit sharing was initiated via the TBGRI-Kani Trust agreement.
Neem Patent Case (India)	Patents on neem-derived products by foreign entities.	**Unethical**, traditional uses were known; patent revoked.
Hoodia and the San People (South Africa)	Appetite suppressant properties used commercially.	Initially **unethical**, later resolved via benefit sharing agreements.

7. Towards a Just and Ethical Future

A just biotechnology framework must transcend legal compliance and actively incorporate **ethical accountability**, ensuring:

- **Recognition of community knowledge systems as valid and scientific**.
- **Long-term partnerships** rather than short-term extraction.
- Integration of **interdisciplinary approaches** combining science, ethics, law, and anthropology.
- Development of **biocultural ethics** that honor the interrelation of biodiversity and culture.

8. Conclusion

Ethical approaches to benefit sharing and indigenous knowledge protection are **not only a matter of fairness but also of sustainability and legitimacy** in biotechnology. As science advances into deeper genomic explorations and synthetic biology, it is imperative that the **voices of indigenous peoples are heard, respected, and integrated** into global innovation systems.

Through **participatory governance, equitable benefit distribution**, and **cultural sensitivity**, biotechnology can become a force for not just discovery, but also for justice and shared prosperity.

Appendices

Appendix I: Key Legal Instruments and Treaties on IPR and Biosafety

This appendix provides a comprehensive list and critical overview of the major international and national legal instruments, treaties, protocols, and frameworks that shape the landscape of Intellectual Property Rights (IPR) and Biosafety in the context of biotechnology. These instruments collectively address issues of innovation protection, equitable benefit sharing, trade, ethical standards, and safety of genetically modified organisms (GMOs).

A. Intellectual Property Rights (IPR) – Legal Instruments and Treaties

1. Paris Convention for the Protection of Industrial Property (1883)

- **Objective:** To establish a common international framework for the protection of industrial property, including patents, trademarks, industrial designs, and geographical indications.
- **Key Provisions:** National treatment, right of priority, independence of patents in different countries.
- **Significance:** One of the first international treaties to harmonize IPR protection across nations.

2. Berne Convention for the Protection of Literary and Artistic Works (1886)

- **Objective:** To protect the rights of creators over their literary and artistic works.
- **Key Provisions:** Automatic protection, moral rights, and exclusive economic rights.
- **Relevance to Biotechnology:** Governs copyright issues in bioinformatics, software used in genome analysis, and databases.

3. Trade-Related Aspects of Intellectual Property Rights (TRIPS) Agreement (1994)

- **Administered by:** World Trade Organization (WTO)

- **Objective:** To establish minimum standards for IPR protection and enforcement among WTO member nations.
- **Key Provisions:**
  - Protection of patents for 20 years,
  - Coverage of biotechnological inventions,
  - Requirements for patentability: novelty, inventive step, and industrial applicability.
- **Relevance:** Compels countries to provide patent protection for microorganisms and microbiological processes, with certain flexibilities for plants and animals.

4. UPOV Convention (International Union for the Protection of New Varieties of Plants) (1961, revised in 1972, 1978, 1991)

- **Objective:** To protect new varieties of plants by granting Plant Breeder's Rights (PBR).
- **Key Concepts:** Distinctness, Uniformity, Stability (DUS) testing.
- **1991 Act Impact:** Stronger PBR, reduced farmers' exemption, introduced EDVs (Essentially Derived Varieties).

5. Budapest Treaty on the International Recognition of the Deposit of Microorganisms for the Purposes of Patent Procedure (1977)

- **Objective:** Facilitates the patenting of microorganisms by recognizing the deposit of samples in designated International Depository Authorities (IDAs).
- **Relevance:** Simplifies procedural aspects of biotechnological patent applications involving living organisms.

6. Patent Cooperation Treaty (PCT) (1970)

- **Objective:** Streamlines the process for filing patents internationally.
- **Key Feature:** A single international application can be filed in multiple countries, deferring national phase entry.
- **Relevance:** Widely used in biotech patent filings to secure inventions globally.

7. Protection of Plant Varieties and Farmers' Rights (PPV&FR) Act, 2001 (India)

- **Objective:** To balance the rights of plant breeders and farmers.
- **Key Provisions:**

   - Registration of new, extant, and farmers' varieties,
   - Benefit-sharing mechanisms,
   - Farmers' rights to save, use, exchange, and sell seeds.
- **Uniqueness:** Recognizes and protects traditional knowledge and informal innovation.

B. Biosafety – Legal Instruments and International Treaties

1. Convention on Biological Diversity (CBD) (1992)

- **Objectives:**
   - Conservation of biological diversity,
   - Sustainable use of its components,
   - Fair and equitable sharing of benefits arising from genetic resources.
- **Relevance to Biosafety:** Recognized the need to ensure safe handling of GMOs and laid the foundation for the Cartagena Protocol.

2. Cartagena Protocol on Biosafety (2000, enforced 2003)

- **A Supplement to:** CBD
- **Objective:** To ensure the safe transfer, handling, and use of LMOs (Living Modified Organisms) that may have adverse effects on biodiversity and human health.
- **Key Provisions:**
   - Advanced Informed Agreement (AIA),
   - Biosafety Clearing House (BCH),
   - Labeling and documentation requirements.
- **Scope:** Applies to transboundary movement of GMOs.

3. Nagoya Protocol on Access and Benefit Sharing (2010)

- **Objective:** Implements fair and equitable sharing of benefits arising from the utilization of genetic resources.
- **Key Provisions:**
   - Prior Informed Consent (PIC),
   - Mutually Agreed Terms (MAT),
   - Compliance mechanisms.
- **Importance:** Enhances protection of indigenous knowledge and biodiversity against biopiracy.

4. Codex Alimentarius Guidelines (by FAO/WHO)

- **Objective:** Establishes international food safety standards, including assessment of GM food.
- **Relevance:** Provides guidance on food risk assessment, allergenicity, and labeling of genetically modified food products.

5. OECD Guidelines for Recombinant DNA Safety Considerations (1986)

- **Objective:** Offers baseline principles for the safe handling of genetically engineered organisms.
- **Use:** Early international reference for containment, risk assessment, and environmental safety.

6. National Biosafety Frameworks (e.g., India's Rules, 1989 under EPA Act)

- **Objective:** To regulate GM research, testing, and release at the national level.
- **Indian Framework Includes:**
  - Institutional Biosafety Committees (IBSC),
  - Review Committee on Genetic Manipulation (RCGM),
  - Genetic Engineering Appraisal Committee (GEAC).

C. Summary Table: Legal Instruments at a Glance

Instrument/ Treaty	Year	Area Covered	Administering Body	Key Relevance
Paris Convention	1883	Industrial Property	WIPO	Patent standards
TRIPS Agreement	1994	Comprehensive IPR	WTO	Global IPR norms
PCT	1970	Patent Filing	WIPO	Global patent application
PPV&FR Act	2001	Plant Varieties	Govt. of India	Farmers' & Breeders' rights
CBD	1992	Biodiversity	UNEP	Genetic resource access
Cartagena Protocol	2000	Biosafety	CBD Secretariat	GMO handling and safety
Nagoya Protocol	2010	Benefit Sharing	CBD Secretariat	Indigenous knowledge
Codex Guidelines	2003+	GM Food Safety	FAO/WHO	Risk evaluation
OECD rDNA Guidelines	1986	Biosafety	OECD	GMO risk containment

Conclusion

This appendix provides an integrated overview of the evolving global and national legal architecture supporting biotechnology, ensuring a delicate balance between innovation, safety, and ethical responsibility. A solid understanding of these legal frameworks is essential for PG students and professionals to navigate the regulatory, research, and commercialization landscapes in biosciences.

Appendix II: Important Indian Acts and Institutions Related to Biotechnology Regulations

India, as a signatory to multiple international treaties and conventions related to biotechnology, biosafety, and intellectual property rights, has developed a comprehensive legal and institutional framework to regulate the growth of biotechnology while ensuring ethical conduct, public safety, and environmental protection. This appendix presents the **key legislations and regulatory institutions** that form the foundation of India's biotechnology governance.

A. Major Indian Acts Relevant to Biotechnology and Biosafety

1. The Environment (Protection) Act, 1986

- **Administering Ministry:** Ministry of Environment, Forest and Climate Change (MoEFCC)
- **Purpose:** Provides the umbrella framework for environmental protection, including biosafety regulations.
- **Key Provisions:**
  - Empowers the central government to take measures for environmental protection,
  - Authorizes formulation of rules for the regulation of GMOs.
- **Associated Rules:***Rules for the Manufacture, Use, Import, Export and Storage of Hazardous Microorganisms/Genetically Engineered Organisms or Cells, 1989.*

2. The Biological Diversity Act, 2002

- **Administering Body:** National Biodiversity Authority (NBA)
- **Objective:** Conservation of biological diversity and equitable sharing of benefits arising from the use of biological resources and associated knowledge.
- **Key Features:**
  - Regulates access to biological resources and associated traditional knowledge,
  - Ensures benefit-sharing with local and indigenous communities,
  - Mandates prior approval for obtaining intellectual property based on Indian biological resources.

3. The Protection of Plant Varieties and Farmers' Rights (PPV&FR) Act, 2001

- **Administering Body:** Protection of Plant Varieties and Farmers' Rights Authority
- **Objective:** Provides intellectual property rights to plant breeders and recognizes the rights of farmers.
- **Salient Features:**
  - Registration of new and extant plant varieties,
  - Farmers' rights to save, use, exchange, and sell farm produce,
  - Provision for benefit sharing and compensation for traditional knowledge holders.

4. The Indian Patents Act, 1970 (Amended in 2005)

- **Administering Body:** Office of the Controller General of Patents, Designs & Trade Marks (CGPDTM)
- **Relevance to Biotechnology:**
  - Allows patents on microbiological processes and products,
  - Excludes patenting of plants and animals (other than microorganisms), and essentially biological processes for the production of plants and animals,
  - Includes Section 3(b), 3(c), and 3(j) to limit patentability on ethical, natural, or biological grounds.

5. The Food Safety and Standards Act, 2006

- **Administering Body:** Food Safety and Standards Authority of India (FSSAI)
- **Objective:** To regulate the safety, manufacture, storage, distribution, and sale of food products, including those derived from genetically modified organisms (GMOs).
- **Relevance:**
  - Ensures labeling, risk assessment, and approval of GM food,
  - Establishes standards for GM food safety in line with Codex guidelines.

6. The Drugs and Cosmetics Act, 1940

- **Administering Body:** Central Drugs Standard Control Organization (CDSCO)

- **Relevance:** Regulates biopharmaceutical products, including recombinant DNA-based drugs, gene therapy products, and vaccines.
- **Key Function:** Provides regulatory oversight for clinical trials, product approval, and post-marketing surveillance.

B. Key Indian Regulatory Institutions Governing Biotechnology

Institution	Administering Ministry	Primary Roles and Functions
Department of Biotechnology (DBT)	Ministry of Science and Technology	Policy formulation, funding R&D, capacity building, biosafety frameworks
Genetic Engineering Appraisal Committee (GEAC)	MoEFCC	Approval of large-scale use, field trials, and commercial release of GMOs
Review Committee on Genetic Manipulation (RCGM)	DBT	Monitors research in genetic engineering, approves confined field trials
Institutional Biosafety Committees (IBSCs)	Respective Research Institutions	Internal monitoring of lab-level recombinant DNA work
National Biodiversity Authority (NBA)	MoEFCC	Access and benefit sharing, protection of biodiversity and traditional knowledge
Food Safety and Standards Authority of India (FSSAI)	Ministry of Health and Family Welfare	Risk assessment and regulation of genetically modified food products
Protection of Plant Varieties and Farmers' Rights Authority (PPV&FRA)	Ministry of Agriculture and Farmers Welfare	Registration and protection of new plant varieties, safeguarding farmers' rights
Central Drugs Standard Control Organization (CDSCO)	Ministry of Health and Family Welfare	Regulation of biotech-based drugs, vaccines, and clinical trials
Office of the CGPDTM	Ministry of Commerce and Industry	Administers patents for biotechnology inventions under Indian Patent Act

C. National Guidelines and Biosafety Frameworks

India has adopted a tiered biosafety assessment system, guided by national protocols harmonized with international standards. Some critical biosafety frameworks include:

- **National Guidelines for Gene Therapy Product Development (2020)** Released by DBT and ICMR for ensuring safety and efficacy of human gene therapy products.

- **Guidelines for the Safety Assessment of Foods Derived from Genetically Engineered Plants (2008)** Outlines compositional analysis, toxicological, and allergenicity assessment of GM foods.
- **Biosafety Guidelines and Regulations (1990, 1994, 1998, 2017)** Periodically updated by DBT for regulation of rDNA research, field trials, and transgenic containment.

D. Summary Table of Key Legal Acts and Institutions

Legal Act/Regulation	Year	Relevance to Biotechnology	Administering Body
Environment (Protection) Act	1986	Biosafety, GMO regulation	MoEFCC
Biological Diversity Act	2002	Access and Benefit Sharing	NBA
PPV&FR Act	2001	Plant Variety Protection	PPV&FRA
Indian Patents Act (Amended)	2005	Patentability of biotech inventions	CGPDTM
Food Safety and Standards Act	2006	GM food safety	FSSAI
Drugs and Cosmetics Act	1940	Biopharmaceutical regulation	CDSCO

Conclusion

India's regulatory framework for biotechnology is robust, multidisciplinary, and evolving, balancing scientific innovation with public safety, environmental conservation, and socio-economic equity. A strong institutional network ensures ethical conduct, legal compliance, and harmonization with global biosafety and IPR norms. For postgraduate scholars, comprehension of these national acts and institutional mechanisms is crucial to navigating the Indian biotech research, development, and commercialization ecosystem.

Appendix III: Model Forms for Patent and Biosafety Applications

Biotechnological research and product development in India are subject to various legal approvals, including **patent filing** and **biosafety clearance**. For clarity and practical understanding, this appendix provides model forms based on the actual templates used by regulatory bodies such as the **Office of the Controller General of Patents, Designs and Trade Marks (CGPDTM)** and the **Department of Biotechnology (DBT)** under the **Ministry of Environment, Forest and Climate Change (MoEFCC)**.

These model forms aim to familiarize students and researchers with the **essential procedural frameworks** to ensure compliance and promote best practices in biotechnology innovation and regulation.

Section A: Model Patent Application Forms (India)

Patent applications in India are governed by the **Indian Patents Act, 1970 (as amended in 2005)**. The **primary forms** required are prescribed under the **Patents Rules, 2003**.

Form 1 – Application for Grant of Patent

Field	Details to be Filled
Name of the Applicant(s)	e.g., Dr. A.B. Sharma
Nationality	Indian
Address for Correspondence	Dept. of Biotechnology, XYZ University, India
Title of the Invention	A Novel Microbial Process for Biofertilizer Production
Type of Application	Ordinary / Convention / PCT National Phase
Declaration	Signature and declaration of true inventorship

Form 2 – Provisional / Complete Specification

Section	Details
Title	Same as in Form 1
Field of Invention	Microbial Biotechnology
Background and Prior Art	Description of existing technologies
Summary of Invention	Brief of novelty, utility, and inventive step
Detailed Description	Step-wise explanation of process or product
Claims	Numbered claims of technical features
Drawings/Diagrams	Labeled figures as necessary
Abstract	<150 words technical summary

Form 3 – Statement and Undertaking

Details of any foreign patent application(s) for the same or substantially the same invention.

Form 5 – Declaration as to Inventorship

Filed with the complete specification, indicating the names of the actual inventors.

Form 18 – Request for Examination

Filed within 48 months of the priority date for the Controller to initiate examination of the application.

Note: These forms are submitted online via the Indian Patent Advanced Search System (InPASS) portal.

Section B: Model Biosafety Approval Forms (India)

For conducting research and field testing involving **genetically engineered organisms (GEOs)** or products derived from them, approvals must be sought from designated committees such as **IBSC**, **RCGM**, and **GEAC**, under the **Rules, 1989 of EPA 1986**.

Model Format for IBSC Approval (Institutional Biosafety Committee)

Field	**Sample Entry**
Name of Principal Investigator	Dr. R. Mehra
Affiliation	Biotechnology Dept., ABC University
Title of Project	Development of Bt-Expressing Tomato Lines
Objective	Expression of insecticidal gene in Solanumlycopersicum
Methodology	Agrobacterium-mediated transformation
Containment Level Required	BSL-2
Duration of Experiment	12 months
Risk Assessment Summary	Low environmental risk; containment measures detailed

Model Format for RCGM Approval (Review Committee on Genetic Manipulation)

Filed after IBSC clearance, especially for **confined field trials**, or when the work exceeds containment limits.

Field	**Sample Entry**
Institution Name	Institute of Agri-Biotech Research
Previous IBSC Approval	Yes (enclose copy)
Transgene Details	Cry1Ac gene under CaMV35S promoter
Vector Backbone	Binary vector pBINAR
Transformation Method	Leaf disc method via *Agrobacterium tumefaciens*
Location for Confined Field Trial	University Farm, Tamil Nadu
Biosafety Measures Enforced	Isolation distance, refuge planting, disposal protocols

Model Format for GEAC Approval (Genetic Engineering Appraisal Committee)

For **large-scale or commercial release** of GE crops or products.

Field	**Details**
Product Name	Insect-resistant BtBrinjal
Proposed Area	Multiple agro-climatic zones across 3 states
Trial Design	Randomized block with control checks
Environmental Risk Data	EIA reports, gene flow data, allergenicity reports
Public Consultation	Summary of consultations and stakeholder feedback

Section C: Key Submission Tips and Best Practices

1. **Accuracy and Completeness:** Ensure every field is filled with correct and updated information.
2. **Clear Descriptions:** Provide precise scientific rationale, risk management, and mitigation strategies.
3. **Proper Attachments:** Include maps, gene constructs, risk assessments, and declaration forms.
4. **Ethical and Legal Compliance:** Ensure the work does not infringe on IPR or violate biodiversity/indigenous rights.
5. **Digital Filing:** Use official portals such as:
   - https://ipindia.gov.in for patents,
   - https://biosafety.icar.gov.in for biosafety approvals.

Conclusion

The inclusion of model forms in this appendix serves to prepare PG students and young researchers for practical engagements with regulatory systems. Understanding procedural documentation ensures **regulatory compliance**, strengthens the **ethical integrity** of biotechnology research, and promotes responsible innovation. As India advances in cutting-edge biosciences, capacity-building through familiarity with such documentation becomes an indispensable component of professional training.

Appendix IV: Case Studies on GMOs, Patents, and Biopiracy

Introduction

Case studies are vital pedagogical tools that translate abstract regulatory, ethical, and scientific principles into concrete, real-life situations. In biotechnology, critical debates around GMOs, intellectual property rights (IPRs), and biopiracy reveal the complexity of balancing innovation, public interest, indigenous rights, and environmental protection.

This appendix highlights **six landmark case studies**, grouped under three core themes:

- **A. Genetically Modified Organisms (GMOs)**
- **B. Patents in Biotechnology**
- **C. Biopiracy and Indigenous Knowledge**

Each case study includes background, key issues, stakeholder perspectives, regulatory or judicial outcomes, and critical lessons.

A. Case Studies on GMOs

Case Study A1: Bt Brinjal Trial and Regulatory Halt in India

Background

Bt Brinjal, a genetically modified eggplant engineered to express the *Cry1Ac* gene from *Bacillus thuringiensis*, was developed by Mahyco in collaboration with public institutions.

Issues

- Biosafety concerns regarding gene flow and non-target effects
- Insufficient long-term toxicology data
- Societal concerns, farmer autonomy, and consumer rights

Outcome

In 2010, after extensive public consultations, the Indian government imposed an indefinite **moratorium** on the commercial release of BtBrinjal.

Lessons Learned

- Transparency and stakeholder engagement are critical
- Public trust in science is mediated by ethics and regulatory credibility

Case Study A2: GM Mustard (Dhara Mustard Hybrid - DMH-11)

Background

Developed by the Centre for Genetic Manipulation of Crop Plants (CGMCP), Delhi University, GM Mustard aimed to improve yield via hybrid technology using *barnase/barstar* genes.

Issues

- Debate over yield claims
- Environmental risks to pollinators
- Allegations of conflict of interest

Outcome

After prolonged review, the **GEAC approved** confined environmental release in 2022, but final approval awaited government consent.

Lessons Learned

- Institutional independence and peer review are essential
- Regulatory decision-making must balance science and sustainability

B. Case Studies on Patents in Biotechnology

Case Study B1: The Neem Patent Controversy

Background

In 1995, the European Patent Office (EPO) granted a patent to W.R. Grace (US) and USDA for a method of neem oil extraction for antifungal use.

Issues

- The technique had long been known in Indian traditional medicine
- Filed without prior informed consent from local communities

Outcome

A challenge by Indian and European NGOs led to the **revocation of the patent** in 2000 by EPO, citing lack of novelty.

Lessons Learned

- Documenting traditional knowledge (e.g., TKDL) is critical
- Ethical IP regimes must respect collective knowledge rights

Case Study B2: Basmati Rice Patent Dispute

Background

In 1997, RiceTec Inc. (US) was granted a US patent (No. 5663484) for "Basmati rice lines and grains."

Issues

- Attempted monopolization of traditional Basmati traits
- Potential impact on Indian rice exports and farmers

Outcome

India contested the claim; eventually, **most claims were overturned** following global protests and scientific scrutiny.

Lessons Learned

- Bioprospecting must not lead to biopiracy
- International cooperation is crucial to defend heritage crops

C. Case Studies on Biopiracy and Indigenous Knowledge

Case Study C1: Turmeric Patent Case

Background

Two US-based researchers from the University of Mississippi Medical Center were granted a US patent (No. 5401504) in 1995 for using turmeric powder to heal wounds.

Issues

- The wound-healing property of turmeric is well-documented in Indian Ayurveda
- No novelty or inventiveness existed

Outcome

The **Council of Scientific and Industrial Research (CSIR), India**, successfully challenged the patent; it was **revoked in 1997**.

Lessons Learned

- Legal instruments must protect cultural heritage
- Scientific documentation of traditional practices empowers defense

Case Study C2: Hoodia and the San People

Background

Hoodia cactus was traditionally used by the **San people** of Southern Africa to suppress appetite. CSIR patented its active compound (*P57*) for anti-obesity drugs.

Issues

- Lack of prior informed consent
- No initial benefit-sharing with indigenous communities

Outcome

Following global criticism, CSIR **entered a benefit-sharing agreement** with the San people in 2003.

Lessons Learned

- Ethical biotechnology respects indigenous sovereignty
- Fair and equitable benefit-sharing must be institutionalized

Conclusion

These case studies reflect the **ethical, legal, and societal crossroads** that modern biotechnology encounters. They highlight the importance of:

- **Informed consent**
- **Respect for traditional knowledge**
- **Transparent regulatory mechanisms**
- **Equitable benefit-sharing**

For postgraduate students, understanding these narratives builds critical perspectives in ethical research and regulatory compliance.

Appendix V: Abbreviations and Acronyms

Abbreviation	Full Form
AIA	Advance Informed Agreement
APEC	Asia-Pacific Economic Cooperation
APHIS	Animal and Plant Health Inspection Service (USA)
ARS	Agricultural Research Service (USDA)
ASEAN	Association of Southeast Asian Nations
BCIL	Biotech Consortium India Limited
Bt	Bacillus thuringiensis
CBD	Convention on Biological Diversity
CBI	Confidential Business Information
CFT	Confined Field Trial
CGMCP	Centre for Genetic Manipulation of Crop Plants
CIET	Centre for Innovation in Education and Technology
CPGR	Commission on Plant Genetic Resources
CSIR	Council of Scientific and Industrial Research
DBT	Department of Biotechnology (India)
DNA	Deoxyribonucleic Acid
DMH-11	Dhara Mustard Hybrid-11
EIA	Environmental Impact Assessment
EOP	End of Process
EPO	European Patent Office
EPA	Environmental Protection Agency (USA)
FAO	Food and Agriculture Organization
FTO	Freedom to Operate
GATT	General Agreement on Tariffs and Trade
GEAC	Genetic Engineering Appraisal Committee
GM	Genetically Modified
GMO	Genetically Modified Organism
GRN	Gene Regulatory Network
GRAS	Generally Recognized as Safe
ICGEB	International Centre for Genetic Engineering and Biotechnology
ICMR	Indian Council of Medical Research
IEEP	Institute for European Environmental Policy
IP	Intellectual Property
IPR	Intellectual Property Rights
ISO	International Organization for Standardization
ISTA	International Seed Testing Association

ITK	Indigenous Traditional Knowledge
MTA	Material Transfer Agreement
MoEFCC	Ministry of Environment, Forest and Climate Change (India)
NAC	National Advisory Committee
NBA	National Biodiversity Authority
NBPGR	National Bureau of Plant Genetic Resources
NBT	New Breeding Techniques
NGM	Non-Genetically Modified
OECD	Organisation for Economic Co-operation and Development
PBR	People's Biodiversity Register
PCR	Polymerase Chain Reaction
PPV&FR	Protection of Plant Varieties and Farmers' Rights
PVP	Plant Variety Protection
RCGM	Review Committee on Genetic Manipulation
RFLP	Restriction Fragment Length Polymorphism
RNA	Ribonucleic Acid
SAU	State Agricultural University
SDN	Site-Directed Nuclease
SOP	Standard Operating Procedure
TKDL	Traditional Knowledge Digital Library
TRIPS	Trade-Related Aspects of Intellectual Property Rights
UNEP	United Nations Environment Programme
UNESCO	United Nations Educational, Scientific and Cultural Organization
USDA	United States Department of Agriculture
UPOV	International Union for the Protection of New Varieties of Plants
WHO	World Health Organization
WIPO	World Intellectual Property Organization
WTO	World Trade Organization

Note to Students

This appendix serves as a **ready-reference tool** to navigate the extensive technical and institutional vocabulary embedded throughout the book. Learners are encouraged to **refer frequently** to this list for clarity and enhanced comprehension of scientific, legal, and policy-related texts in biotechnology.

Appendix VI: Glossary of Terms

Advance Informed Agreement (AIA)

A procedure under the Cartagena Protocol that ensures countries are informed about the first intentional transboundary movement of living modified organisms (LMOs) and consent is granted before such movement occurs.

Allergenicity

The potential of a substance, particularly a genetically modified protein, to cause allergic reactions in sensitive individuals.

Bacillus thuringiensis (Bt)

A naturally occurring soil bacterium that produces insecticidal toxins, widely used in GM crops for pest resistance.

Benefit Sharing

A principle under the Convention on Biological Diversity (CBD) that promotes fair and equitable sharing of benefits arising from the utilization of genetic resources with the source communities or countries.

Biopiracy

Unauthorized use or patenting of biological resources or associated traditional knowledge without adequate compensation or acknowledgment of the indigenous communities.

Biotechnology

The application of scientific and engineering principles to the processing of materials by biological agents to provide goods and services.

Biosafety

The set of measures and protocols adopted to ensure safe handling, transfer, and use of genetically modified organisms, minimizing risks to health and the environment.

Cartagena Protocol on Biosafety

An international agreement that governs the safe transfer, handling, and use of living modified organisms resulting from modern biotechnology.

Confined Field Trials (CFT)

Experimental cultivation of genetically modified crops in a controlled environment under restricted and monitored conditions to study their performance and assess risks.

Cry Genes

Genes derived from *Bacillus thuringiensis* that code for crystal proteins toxic to specific insects, widely used in transgenic crops for pest resistance.

Environmental Risk Assessment (ERA)

A systematic process to identify, evaluate, and mitigate the potential adverse environmental effects of genetically modified organisms.

Familiarity

A concept in risk assessment that relies on existing knowledge and prior experience with an organism or its traits to inform safety evaluations.

Gene Flow

The movement or transfer of genetic material from genetically modified organisms to non-GM populations or wild relatives through pollen or seed dispersal.

Gene Regulatory Network (GRN)

A collection of regulatory interactions among genes and transcription factors that control gene expression and coordinate biological processes.

Genetic Engineering

The deliberate modification of an organism's genetic material using recombinant DNA technology to introduce new traits or functions.

Genetic Use Restriction Technologies (GURTs)

Biotechnological methods developed to control the propagation of genetically modified crops by rendering second-generation seeds sterile.

Genetically Modified Organisms (GMOs)

Organisms whose genetic material has been altered in a way that does not occur naturally through mating or natural recombination.

Horizontal Gene Transfer

The transfer of genetic material between organisms in a manner other than traditional reproduction, often involving microorganisms.

Indigenous Knowledge (IK)

The traditional knowledge, innovations, and practices of local communities developed over generations, often related to biodiversity and agriculture.

Labeling (of GMOs)

The regulatory requirement to provide information on food products derived from genetically modified organisms to ensure consumer awareness and choice.

Living Modified Organisms (LMOs)

Any living organism that possesses a novel combination of genetic material obtained through modern biotechnology.

Material Transfer Agreement (MTA)

A legal document governing the transfer of tangible biological materials between organizations, including rights and obligations.

Molecular Characterization

A detailed analysis of the molecular composition of a genetically modified organism, including the inserted gene(s), expression levels, and genetic stability.

Nagoya Protocol

A supplementary agreement to the CBD that provides a legal framework for access to genetic resources and fair benefit sharing.

Patent

An exclusive legal right granted for an invention that is novel, involves an inventive step, and is industrially applicable.

Phenotypic Assessment

The evaluation of observable traits in genetically modified organisms to determine the expression and stability of introduced genes.

Post-release Monitoring

The systematic observation of genetically modified organisms after they are released into the environment to detect and manage unforeseen effects.

Risk Assessment

A science-based process used to estimate the potential adverse effects of genetically modified organisms on human health and the environment.

Site-Directed Nuclease (SDN)

Genome editing tools such as CRISPR, TALENs, or ZFNs used to introduce precise changes in an organism's DNA sequence.

Standard Operating Procedure (SOP)

A detailed, written instruction designed to achieve uniformity in the performance of specific functions, particularly in biosafety and laboratory protocols.

Substantial Equivalence

A principle that compares a genetically modified organism with its non-modified counterpart to identify meaningful differences in safety assessments.

Toxicity

The degree to which a substance can harm organisms, evaluated during safety assessments of GMOs for food and environmental use.

Traceability

The ability to track the origin, history, and movement of a genetically modified organism or its derived product through all stages of production and distribution.

Transgenic Plant

A plant that contains a gene or genes which have been artificially inserted instead of the plant acquiring them through pollination.

Traditional Knowledge Digital Library (TKDL)

An Indian initiative that digitally records indigenous knowledge related to medicine and agriculture to prevent biopiracy and wrongful patenting.

UPOV Convention The International Union for the Protection of New Varieties of Plants, promoting intellectual property rights for plant breeders.

WTO TRIPS Agreement

The World Trade Organization agreement on Trade-Related Aspects of Intellectual Property Rights, setting minimum standards for IPR regulation among member countries.

Note to Students

This glossary serves as a **quick reference tool** for understanding complex and domain-specific terms. It reinforces conceptual clarity, particularly in preparation for **academic assessments, research interpretation, and policy comprehension**.

Appendix VII: Web Resources and Government Portals

This section lists important national and international websites and portals that provide authoritative information, databases, policy documents, updates, and application procedures relevant to Intellectual Property Rights (IPR), biosafety, biotechnology regulation, bioethics, and environmental concerns. Students, researchers, and professionals are encouraged to regularly consult these digital platforms for current developments and primary documents.

Intellectual Property Rights (IPR)

1. World Intellectual Property Organization (WIPO)

   https://www.wipo.int

   → Provides global IP treaties, patent databases (PATENTSCOPE), and IPR policy documents.

2. **Intellectual Property India (CGPDTM)**

   https://ipindia.gov.in

   → Official portal of India's Controller General of Patents, Designs and Trade Marks. Patent search, filing forms, IPR laws, and updates are available.

3. **United States Patent and Trademark Office (USPTO)**

   https://www.uspto.gov

   → Search and access US patent records and legal information.

4. **European Patent Office (EPO)**

   https://www.epo.org

   → Offers global patent databases (Espacenet), examination procedures, and training modules.

Biosafety and GMO Regulation

5. Cartagena Protocol on Biosafety – Convention on Biological **Diversity (CBD)**

   https://bch.cbd.int/protocol

   → Official repository for international biosafety decisions, LMOs, risk assessments, and national reports.

6. **Genetic Engineering Appraisal Committee**

   GEAC), MoEFCC – Indiahttp://geacindia.gov.in

   → India's apex body for granting approvals of GMOs. Includes biosafety guidelines, application formats, and CFT protocols.

7. **Biosafety Clearing-House (BCH)**

   https://bch.cbd.int

   → Provides access to scientific, technical, environmental, and legal information about LMOs.

8. **Department of Biotechnology (DBT), India**

   https://dbtindia.gov.in

   → Publishes biosafety guidelines, notifications, and regulatory documents for GM research.

9. **Food Safety and Standards Authority of India (FSSAI)**

   https://fssai.gov.in

   → Regulatory body for food safety assessment including genetically modified food products.

Ethics, Bioethics, and Benefit Sharing

10. UNESCO Bioethics Programme

    https://en.unesco.org/themes/ethics-science-and-technology

    → Ethical frameworks and declarations on human rights and bioethics.

11. **Indian Council of Medical Research (ICMR)**

    https://ethics.icmr.org.in

    → National Ethical Guidelines for biomedical and health research involving human participants.

12. **National Biodiversity Authority (NBA), India**

    https://nbaindia.org

    → Information on biodiversity conservation, access and benefit-sharing (ABS), and traditional knowledge protection.

13. **Convention on Biological Diversity (CBD)**

    https://www.cbd.int

    → Global platform for biodiversity conservation, Nagoya Protocol, and traditional knowledge rights.

Patent Filing and Legal Forms

14. **e-Governance Portal for Patent Filing (India)**

    https://ipindiaservices.gov.in

    → Online portal to file patents, trademarks, and access legal forms under CGPDTM.

15. **Patent Cooperation Treaty (PCT) – WIPO**

    https://www.wipo.int/pct/en/

    → Access to international patent application system and procedural guidelines.

Databases and Tools

16. **National Centre for Biotechnology Information (NCBI)**

    https://www.ncbi.nlm.nih.gov

    → Key bioinformatics resources, genomic databases, and molecular tools.

17. **India Biodiversity Portal**

    https://indiabiodiversity.org

    → Open-access portal for biodiversity documentation and indigenous knowledge.

18. **INFLIBNET and Shodhganga**

    https://shodhganga.inflibnet.ac.in

    → Repository of Indian theses, dissertations, and scholarly research.

Note to Students and Researchers:

Always refer to the latest notifications, revised guidelines, and government circulars from the official portals. These sources are authoritative and should be cited in academic, legal, and policy-related writings.

Appendix VIII: Suggested Readings

Part I: Intellectual Property Rights and Legal Framework

1. Ganguli, P. (2001). *Intellectual property rights: Unleashing the knowledge economy*. Tata McGraw-Hill.
2. World Intellectual Property Organization (WIPO). (2020). *Understanding intellectual property*. https://www.wipo.int/edocs/pubdocs/en/intproperty/895/wipo_pub_895.pdf
3. Krattiger, A., Mahoney, R. T., Nelsen, L., Thomson, J. A., Bennett, A. B., Satyanarayana, K., & Graff, G. D. (2007). *Intellectual property management in health and agricultural innovation: A handbook of best practices*. MIHR & PIPRA.
4. Government of India. (2005). *The Patents Act, 1970 (as amended)*. Controller General of Patents, Designs and Trademarks. https://ipindia.gov.in/writereaddata/Portal/ev/sections/ps1.html
5. Ministry of Agriculture, Government of India. (2001). *The Protection of Plant Varieties and Farmers' Rights (PPV&FR) Act*.
6. World Trade Organization (WTO). (1995). *Agreement on Trade-Related Aspects of Intellectual Property Rights (TRIPS)*. https://www.wto.org/english/tratop_e/trips_e/trips_e.htm

Part II: Biosafety and Regulatory Frameworks

1. World Health Organization (WHO). (2004). *Laboratory biosafety manual* (3rd ed.). WHO Press.
2. Department of Biotechnology (DBT), Government of India. (1990). *Recombinant DNA safety guidelines*. Ministry of Science and Technology.
3. Secretariat of the Convention on Biological Diversity.(2000). *Cartagena Protocol on Biosafety to the Convention on Biological Diversity: Text and annexes*. https://bch.cbd.int/protocol/
4. Ministry of Health and Family Welfare, Government of India. (2020). *Biosafety guidelines for containment facilities and laboratory practices*. National Institute of Virology.
5. Food and Agriculture Organization (FAO). (2001). *Manual on biosafety of genetically modified organisms*. FAO Publications.
6. Tzotzos, G. T., Head, G., & Hull, R. (2009). *Genetically modified plants: Assessing safety and managing risk*. Elsevier.

Part III: Risk and Safety Assessment of Transgenics

1. Organisation for Economic Co-operation and Development (OECD). (1993). *Safety evaluation of foods derived by modern biotechnology: Concepts and principles*. OECD Publishing.
2. European Food Safety Authority (EFSA). (2011). *Guidance for risk assessment of food and feed from genetically modified plants*. EFSA Journal, 9(5), 2150. https://doi.org/10.2903/j.efsa.2011.2150
3. National Research Council (NRC). (2004). *Biological confinement of genetically engineered organisms*. The National Academies Press. https://doi.org/10.17226/10880
4. FAO/WHO. (2001). *Evaluation of allergenicity of genetically modified foods*. http://www.fao.org/docrep/003/x6871e/x6871e00.htm
5. Department of Biotechnology, Government of India. (2008). *Regulatory guidelines for GM crops*. Ministry of Science and Technology. https://dbtindia.gov.in/
6. Ehlers, J. D., & Hall, A. E. (2007). *Monitoring and post-release detection of GMOs*. Springer.

Part IV: Bioethics and Societal Issues in Biotechnology

1. Beauchamp, T. L., & Childress, J. F. (2019). *Principles of biomedical ethics* (8th ed.). Oxford University Press.
2. United Nations Educational, Scientific and Cultural Organization (UNESCO).(2005). *Universal Declaration on Bioethics and Human Rights*. https://unesdoc.unesco.org/ark:/48223/pf0000146180
3. Convention on Biological Diversity (CBD) Secretariat.(2011). *Nagoya Protocol on Access and Benefit-sharing*. https://www.cbd.int/abs/
4. Posey, D. A., & Dutfield, G. (1996). *Beyond intellectual property: Toward traditional resource rights for indigenous peoples and local communities*. International Development Research Centre (IDRC).
5. Choudhary, B., & Gaur, K. (2009). *Genetically modified crops: Myth and reality*. ISAAA Brief No. 40.
6. Swaminathan, M. S. (2001). Ethics and equity in genetic resource management. *Current Science*, 80(4), 495–500.

Appendix IX: Review Questions

Part I: Intellectual Property Rights and Legal Framework

1. Explain the evolution of Intellectual Property Rights (IPRs) with special reference to their emergence in modern biotechnology.
2. Discuss the role and importance of IPR in the Indian biotechnology sector. What are the current challenges and opportunities?
3. Compare and contrast different types of intellectual property rights with suitable biotechnological examples.
4. What are the essential criteria for patentability in biotechnology? Illustrate with examples.
5. Critically evaluate the controversies surrounding patenting of genes, living organisms, and biotechnological processes.
6. Describe the process of filing a patent in India and outline the major legal and documentary requirements.
7. What is the Patent Cooperation Treaty (PCT)? How does it facilitate international patenting in biotechnology?
8. Differentiate between patents, copyrights, and plant variety protections in the context of biotech innovation.
9. Discuss the provisions and significance of the Protection of Plant Varieties and Farmers' Rights (PPV&FR) Act.
10. Assess the impact of the TRIPS Agreement on global IPR norms, particularly in relation to biotechnology and germplasm movement.

Part II: Biosafety and Regulatory Frameworks

1. Define biosafety and biohazards. How are biohazards classified in biotechnology?
2. Elucidate the scope and importance of biosafety in laboratories and environmental contexts.
3. Describe the major principles of biosafety and how they are applied to contain risks in biotech labs.
4. Outline the steps involved in conducting a biosafety risk assessment in research laboratories.
5. Explain the biosafety levels (BSL-1 to BSL-4) and discuss their relevance in handling GMOs.
6. What are the standard practices for handling, storage, and disposal of biohazardous materials?

7. Analyze a case study of a biosafety lapse and suggest how it could have been prevented.
8. Describe the objectives and implementation mechanisms of the Cartagena Protocol on Biosafety.
9. Discuss the Indian biosafety regulatory system. What are the roles of key authorities like RCGM, GEAC, and IBSC?
10. Compare international biosafety policies and explain how institutional biosafety committees function within this framework.

Part III: Risk and Safety Assessment of Transgenics

1. Discuss the major environmental and food safety concerns associated with transgenic plants.
2. What are the public concerns surrounding GMOs, and how can these be scientifically addressed?
3. Explain the biosafety concerns related to gene flow, weediness, and impacts on non-target organisms in transgenic crops.
4. Describe the conceptual framework and principles involved in risk assessment of genetically modified (GM) crops.
5. Define "familiarity" and "substantial equivalence" in GMO risk evaluation and provide examples.
6. How are toxicity and allergenicity assessed during the safety evaluation of GM crops?
7. Evaluate the environmental impact assessment of GM crops using specific case examples.
8. What strategies are employed for pre- and post-release monitoring of transgenic organisms?
9. Describe the sampling and analytical techniques used for detection and traceability of transgenics.
10. Discuss the global regulatory requirements for labeling and traceability of genetically modified products.

Part IV: Bioethics and Societal Issues in Biotechnology – Review Questions

1. What are the major ethical considerations in modern biotechnology research and development?
2. Discuss the role of religious, social, and cultural perspectives in shaping bioethical decisions.

3. How do environmental ethics influence regulatory decisions concerning biodiversity and GMOs?
4. What are the components and importance of ethical guidelines for genetic engineering practices?
5. Explain the need and protocols for biosafety research trials and confined field trials (CFTs).
6. What are Standard Operating Procedures (SOPs), and how do they contribute to biosafety in field trials?
7. Define biopiracy. Provide notable examples and analyze their socio-economic consequences.
8. Examine the role of international treaties like the CBD and Nagoya Protocol in preventing biopiracy.
9. Discuss the concept of benefit sharing and how it aligns with ethical norms and indigenous rights.
10. How can responsible biotech innovation ensure equity, transparency, and societal acceptance?

Appendix X: Multiple Choice Questions (MCQs)

Part I: Intellectual Property Rights and Legal Framework

1. The term "Intellectual Property Rights" broadly refers to:
   a) Physical ownership of land and materials
   b) Rights given to persons for creations of their minds
   c) Government-imposed restrictions on inventions
   d) Biological patenting in agriculture
2. Which international agreement was pivotal in standardizing global IPR rules?
   a) GATT **b) TRIPS**
   c) WIPO d) CBD
3. Patents generally offer protection for a period of:
   a) 5 years b) 10 years
   c) 20 years d) Lifetime of the inventor
4. The Indian act that protects the rights of plant breeders and farmers is:
   a) IPR Act **b) PPV&FR Act**
   c) Biodiversity Act d) Patent Act, 1970
5. A trade secret primarily differs from a patent because it:
   a) Requires government registration
   b) Is disclosed to the public
   c) Is protected without disclosure
   d) Applies only to agriculture
6. Which of the following is *not* patentable in India?
   a) Engineered microorganisms
   b) Novel chemical processes
   c) Genetically modified crops
   d) Discoveries of natural genes
7. The term "prior art" refers to:
   a) Existing knowledge or inventions relevant to a new claim
   b) A legal document signed before filing a patent
   c) Artistic designs submitted before copyright
   d) An art form used in biotechnology

8. Industrial designs protect:
    a) Software algorithms
    b) Aesthetic features of products
    c) Technical innovation in engineering
    d) Genetic sequences
9. Which organization administers the Patent Cooperation Treaty (PCT)?
    a) WIPO b) WTO
    c) TRIPS d) UPOV
10. Which one of the following can be copyrighted?
    a) Software code b) Gene sequence
    c) Biotechnological process d) Enzyme structures
11. The term "bioprospecting" is best described as:
    a) Stealing indigenous knowledge
    b) Exploration of biological material for commercial gain
    c) Identification of medicinal plants only
    d) Patent infringement by MNCs
12. Which Indian institute is primarily responsible for patent administration?
    a) CSIR **b) Indian Patent Office**
    c) DBT d) PPV&FR Authority
13. "Substantial equivalence" is a term primarily associated with:
    a) Comparison of GM crops with conventional ones
    b) Equal licensing fees
    c) Bioethics in animal research
    d) Farmers' rights under WTO
14. The main criterion for granting a patent does *not* include:
    a) Novelty b) Inventive step
    c) Duration of research d) Industrial applicability
15. The TRIPS Agreement was signed under the umbrella of:
    a) WTO b) FAO
    c) CBD d) UNFCCC
16. Copyright laws do not protect:
    a) Ideas b) Poems
    c) Music compositions d) Original literary works

17. Which Indian law protects geographical indications?
    a) Trademark Act
    b) GI of Goods (Registration and Protection) Act, 1999
    c) IPR Act
    d) Design Act
18. "Plant Breeder's Rights" are recognized under:
    a) The Patent Act
    b) The PPV&FR Act
    c) The Copyright Act
    d) The GI Act
19. Which body governs industrial design registration in India?
    a) Controller General of Patents, Designs and Trademarks
    b) WIPO
    c) CBD Secretariat
    d) NBPGR
20. Compulsory licensing is allowed under:
    a) WTO patent exemptions
    b) PPV&FR Act
    c) Indian Patent Act
    d) TRIPS Section 3
21. Which of the following is an international organization dealing specifically with IPR?
    a) WTO
    b) WIPO
    c) FAO
    d) WHO
22. Gene patenting is:
    a) Permitted unconditionally in India
    b) Not possible anywhere
    c) Controversial and regulated globally
    d) Mandated by PPV&FR Act
23. The Budapest Treaty deals with:
    a) Cross-border IPR enforcement
    b) Bioprospecting ethics
    c) Microorganism deposit for patent procedure
    d) Patent fee regulation

24. A "patent thicket" is:
    a) A cluster of plants under legal rights
    b) A dense web of overlapping patents
    c) An act of piracy in biotech
    d) A GMOs restriction in India
25. TRIPS was enforced globally in:
    a) 1991 **b) 1995**
    c) 2000 d) 2005
26. The legal term for unauthorized commercial use of a patented invention is:
    a) Compulsory licensing **b) Infringement**
    c) Assignment d) Disclosure
27. The Indian Patent Act was amended in:
    a) 2005 b) 1990
    c) 2015 d) 2010
28. The main aim of PPV&FR Act is:
    a) Trade regulation
    b) Protecting both plant breeders and farmers
    c) Promoting TRIPS
    d) Limiting gene flow
29. Which of the following cannot be protected under IPR?
    a) Naturally occurring bacteria b) Transgenic mice
    c) Modified enzymes d) Recombinant proteins
30. An exclusive legal right to use an invention is called:
    a) Copyright **b) Patent**
    c) Trademark d) GI

Part II: Biosafety and Regulatory Frameworks

1. The primary purpose of biosafety regulations is to:
    a) Protect human health and the environment from risks of biotechnology
    b) Promote rapid release of GMOs
    c) Ensure exclusive patent rights to inventors
    d) Substitute ethics in research

2. The Cartagena Protocol primarily addresses:
   a) Climate change legislation
   b) Transboundary movement of living modified organisms (LMOs)
   c) Global patent harmonization
   d) Traditional knowledge
3. Institutional Biosafety Committees (IBSCs) in India operate under:
   a) Department of Biotechnology (DBT)
   b) ICMR
   c) FAO
   d) CSIR
4. Which body is responsible for approval of large-scale use of GMOs in India?
   a) RCGM b) IBSC
   c) GEAC d) MOEFCC
5. Biosafety Level 4 (BSL-4) laboratories are designed to handle:
   a) Bacterial vaccines
   b) Agricultural pests
   c) Plant growth promoters
   d) Dangerous and exotic pathogens
6. Risk assessment in biosafety involves:
   a) Marketing GMOs for public use
   b) Legalizing all biotech crops
   c) Evaluating potential adverse effects of GMOs
   d) Ignoring unintended gene flow
7. The Recombinant DNA Safety Guidelines in India were introduced in:
   a) 1987 b) 1999
   c) 1990 d) 2002
8. The term "confinement strategy" in biosafety refers to:
   a) Arresting researchers violating protocol
   b) Minimizing yield loss
   c) Reducing the spread of transgenes into the environment
   d) Enhancing nitrogen fixation

9. The apex body for approving biotech product commercialization in India is:

   a) DBT **b) GEAC**

   c) RCGM d) ICAR

10. "Substantial equivalence" in biosafety means:

    a) GM product is comparable in composition to its non-GM counterpart

    b) Patent equivalence in global markets

    c) Legal harmonization of biotech laws

    d) Gene editing creates superior equivalents

11. The term "unintended effects" in GMO evaluation refers to:

    a) Unexpected consequences from genetic modification

    b) Intentional overexpression

    c) Unpatented traits

    d) Commercial restrictions

12. Which is the correct chronological order of risk assessment steps?

    a) Exposure → Management → Communication → Characterization

    b) Hazard ID → Exposure → Characterization → Marketing

    c) Hazard ID → Exposure → Risk characterization → Risk management

    d) Benefit-cost → Hazard → Testing

13. The Cartagena Protocol is legally binding for:

    a) Private seed companies

    b) Parties that have ratified the protocol

    c) Public scientists only

    d) All signatories of CBD regardless of ratification

14. Biosafety Level 1 labs handle:

    a) Ebola virus

    b) Well-characterized, non-pathogenic microbes

    c) GM crops

    d) Antibiotic-resistant fungi

15. The purpose of the Biological Diversity Act, 2002 is to:
    a) Promote gene editing in agriculture
    b) Conserve biodiversity and regulate access to bio-resources
    c) Approve GMOs for export
    d) Train students in molecular biology
16. Risk communication in biosafety means:
    a) Hiding risk data from public
    b) Transferring genes safely
    c) Ensuring lab-to-lab shipment of samples
    d) Transparent information-sharing among stakeholders
17. The "trigger" for biosafety review of a GMO is typically:
    a) Public protest
    b) Farmer complaint
    c) Intended release into the environment
    d) Completion of transformation
18. RCGM stands for:
    a) Review Committee on Genetic Manipulation
    b) Regulatory Council of GMOs
    c) Risk Communication and GMO Monitoring
    d) Revised Code of Gene Management
19. GEAC functions under the:
    a) Ministry of Science and Technology
    b) Indian Council of Medical Research
    c) Ministry of Environment, Forest and Climate Change (MoEFCC)
    d) Department of Agricultural Research
20. The term "event-based approval" in GM crops refers to:
    a) Temporary approvals based on season
    b) Approval linked to specific transformation events
    c) Randomized testing protocol
    d) Social event-based popularization

21. Containment protocols in a biosafety lab are necessary to:
    a) Improve transformation efficiency
    b) Prevent unintentional exposure to recombinant organisms
    c) Increase GMO shelf life
    d) Avoid legal patent disputes
22. The term "gene flow" is critical in biosafety because it:
    a) Improves biodiversity
    b) Aids in cross-pollination
    c) May spread transgenes to wild relatives or non-GM crops
    d) Is used to enhance hybrid vigor
23. Labeling of GM products is necessary to ensure:
    a) Consumer choice and traceability
    b) Improved gene expression
    c) Legal trademark enforcement
    d) Confidentiality of the developer
24. Which body is involved in regulating transgenic research at the laboratory level?
    a) GEAC
    b) Institutional Biosafety Committee (IBSC)
    c) IPR Cell
    d) WHO
25. Biosafety issues of gene editing are:
    a) Negligible due to precision
    b) Approved without regulation
    c) Subject to similar scrutiny as transgenic technologies
    d) Limited to therapeutic use only
26. A "molecular characterization report" is essential in biosafety to:
    a) Advertise GM traits
    b) Apply for trademarks
    c) Describe inserted DNA and its integration site
    d) Ensure patentability

27. Guidelines for conduct of confined field trials in India are issued by:
    a) Department of Biotechnology (DBT)
    b) ICAR
    c) CSIR
    d) GEAC
28. The term "risk management" in biosafety implies:
    a) Profit planning in seed trade
    b) Breeding for drought tolerance
    c) Suppression of data on GMO risks
    d) Mitigating or controlling identified risks
29. Biosafety laws are necessary in biotechnology to:
    a) Hinder research
    b) Promote only foreign GMOs
    c) Ensure responsible research and release of biotech products
    d) Prevent farmer innovation
30. In Indian regulatory system, confined field trials require approval from:
    a) ICAR and CSIR jointly
    b) Review Committee on Genetic Manipulation (RCGM)
    c) IBSC
    d) National Biodiversity Authority

Part III: Risk and Safety Assessment of Transgenics

1. The term "gene flow" in GM crops refers to:
    a) Movement of genes during mitosis
    b) RNA transcription in nucleus
    c) Transfer of transgenes to non-GM organisms or wild relatives
    d) Flow of genetic material from root to shoot
2. Which of the following is a key environmental concern of GMOs?
    a) Enhanced vitamin content
    b) Impact on non-target organisms
    c) Gene silencing
    d) Biodegradation of transgene

3. Substantial equivalence is used to:
   a) Compare GM crops to conventional counterparts for risk evaluation
   b) Determine molecular weight of transgenes
   c) Quantify protein expression
   d) Classify antibiotics
4. A transgenic crop showing herbicide resistance may lead to:
   a) Reduced pesticide cost
   b) Evolution of herbicide-resistant weeds
   c) Delayed flowering
   d) Enhanced nutritional quality
5. The allergenicity of a transgenic food is primarily assessed by:
   a) Field yield trials
   b) Morphological analysis
   c) In vitro digestibility and sequence homology tests
   d) Gene knockout studies
6. Public opposition to GM crops is often rooted in:
   a) Patent regulations
   b) Breeding strategies
   c) Perceived risks and lack of labeling transparency
   d) Controlled pollination
7. Which one is a direct method of transgene detection in monitoring programs?
   a) PCR-based molecular analysis
   b) ELISA strip test
   c) Spectroscopy
   d) Phytochemical screening
8. "Weediness" in biosafety assessment indicates:
   a) Higher resistance to drought
   b) Ability of GM crop to survive outside cultivation and become invasive
   c) Delayed senescence
   d) Gene silencing in hybrids

9. The first step in any GM crop risk assessment is:
   a) Market survey
   b) Public hearing
   c) Hazard identification
   d) Trait evaluation
10. Which of the following indicates a **post-release** monitoring strategy?
    a) Tracking the transgene expression over several generations in the environment
    b) Genomic alignment
    c) Promoter replacement
    d) Cry protein assay in vitro
11. Labeling of GM foods is essential to:
    a) Promote GMO sales
    b) Provide consumer choice and support traceability
    c) Ensure yield uniformity
    d) Comply with trade tariffs
12. The concept of familiarity in biosafety refers to:
    a) Prior scientific knowledge of crop traits and its ecosystem interactions
    b) Familiar genes in public databases
    c) Homology with non-GM sequences
    d) Known trade names
13. Gene transfer from transgenics to soil microbes is called:
    a) Vertical gene transfer
    b) Horizontal gene transfer
    c) Signal transduction
    d) Somatic recombination
14. A GM crop is considered "substantially equivalent" when:
    a) Its nutritional, toxicological, and compositional profiles match the non-GM counterpart
    b) It has identical genetic sequences
    c) Its color and size match wild types
    d) Gene promoters are from the same species
15. Which of the following tools is used for quantitative detection of GM DNA?
    a) DNA barcoding
    b) Real-Time PCR (qPCR)
    c) Immunoblotting
    d) Restriction digestion

16. A critical parameter in environmental safety assessment of GMOs is:
    a) Commercial value
    b) Storage stability
    c) Promoter strength
    d) Potential impact on biodiversity
17. Sampling for GMO detection must be:
    a) Subjective and infrequent
    b) Representative, statistically valid, and contamination-free
    c) Based on farmer preference
    d) Done only during flowering
18. Which technique can be used to detect transgenic proteins in food?
    a) ELISA
    b) Southern blot
    c) RAPD
    d) RT-PCR
19. The principle of traceability in GM regulation ensures:
    a) Marketing under trade names
    b) Seed germination guarantees
    c) Tracking and identification of GM products at every stage
    d) Pest control in the field
20. A possible outcome of transgene escape is:
    a) Reduction in photosynthesis
    b) Loss of soil fertility
    c) Creation of superweeds
    d) Enhanced root formation
21. Which of the following is least likely a public concern about GMOs?
    a) Allergenicity
    b) Increase in test weight
    c) Ethical implications
    d) Labeling transparency
22. An example of a non-target organism potentially affected by GM crops is:
    a) Monarch butterfly
    b) Bt gene
    c) Chloroplast
    d) Soil texture
23. The allergenicity assessment includes:
    a) Immunization of plants
    b) Feeding trials in animals only
    c) Serum screening from allergic individuals and digestibility assays
    d) Herbicide spraying tests

24. Which of these is required for pre-release risk assessment?
    a) Molecular characterization and expression profiling
    b) Export permit
    c) Seed certification
    d) Farmer adoption survey
25. The "event-specific" test in GMO detection targets:
    a) Unique DNA sequence at the insertion site
    b) Plasmid origin
    c) Total genome
    d) Promoter elements only
26. A genetically modified plant might affect food safety due to:
    a) Unexpected production of new proteins or toxins
    b) Pollen viability
    c) Gene silencing in hybrids
    d) Root-shoot ratio
27. Sub-chronic toxicity studies in GM food assessment are conducted in:
    a) Plants
    b) Humans
    c) Laboratory animals like rats or mice
    d) Fungi
28. Sampling strategy for GMO detection should follow:
    a) Farmer-based identification
    b) Statistically random and stratified design
    c) Weather-based scheduling
    d) Visual selection
29. Which global organization provides guidelines for GMO food safety evaluation?
    a) WTO
    b) Codex Alimentarius Commission (FAO/WHO)
    c) UNEP
    d) ICRISAT

30. The primary goal of transgenic detection methods is to:
    a) Identify non-GM crops
    b) Confirm the presence, quantity, and regulatory compliance of GMOs
    c) Detect yield variation
    d) Compare hybrid varieties

Part IV: Bioethics and Societal Issues in Biotechnology

1. Which of the following is a core objective of confined field trials (CFTs)?
    a) To assess biosafety under controlled environmental conditions
    b) To maximize production yield
    c) To bypass regulatory evaluations
    d) To allow open commercial release
2. The Nagoya Protocol focuses on:
    a) Enhancing lab biosafety
    b) Access to genetic resources and equitable benefit-sharing
    c) Agricultural trade reform
    d) Food allergen testing
3. Ethical research in biotechnology mandates:
    a) Informed consent and transparency in experimentation
    b) Profit-driven approaches
    c) Secrecy in data reporting
    d) Rapid commercial rollout
4. The main purpose of a Standard Operating Procedure (SOP) during GMO field trials is to:
    a) Assist farmers in pesticide application
    b) Ensure repeatable and safe conduct of experiments
    c) Improve economic feasibility
    d) Optimize nutrient efficiency
5. An example of biopiracy is:
    a) Gene editing in mice
    b) Patenting traditional turmeric use without acknowledging its origin
    c) Producing transgenic rice with high zinc
    d) Mutagenesis breeding

6. Which international treaty first introduced legally binding commitments for biodiversity protection?
   a) Convention on Biological Diversity (CBD)
   b) Cartagena Protocol
   c) ITPGRFA
   d) Kyoto Protocol
7. The term "bioethics" in biotechnology primarily deals with:
   a) Moral implications of scientific practices involving life forms
   b) Weather prediction
   c) Pesticide residue testing
   d) Market ethics
8. One of the key ethical concerns regarding genetic modification is:
   a) Loss of leaf area
   b) Unintended consequences to biodiversity
   c) Pollination time
   d) Germination rate
9. The regulatory body in India that oversees GMO field trials is:
   a) Review Committee on Genetic Manipulation (RCGM)
   b) CSIR
   c) ICAR
   d) MoFPI
10. "Environmental ethics" in GM technology refers to:
    a) Water-saving farming
    b) Crop quality enhancement
    c) Responsible treatment of ecosystems and biodiversity
    d) Food colorant testing
11. Religious and cultural objections to GMOs often stem from:
    a) Tax burdens
    b) Perceived unnatural tampering of life
    c) Nutrient quality
    d) Herbicide residues

12. Which article of the CBD emphasizes benefit sharing from the use of biological resources?
    a) Article 15
    b) Article 3
    c) Article 8(j)
    d) Article 10
13. The principle of "prior informed consent" (PIC) is critical in:
    a) Lab cloning
    b) Soil testing
    c) Accessing and using traditional knowledge or genetic resources
    d) Market launch of hybrids
14. The Cartagena Protocol primarily governs:
    a) Organic farming practices
    b) Transboundary movement and safe handling of living modified organisms
    c) Antibiotic residue in milk
    d) Rainwater harvesting
15. SOPs in field trials should include:
    a) Financial estimates
    b) Risk mitigation steps, containment procedures, and data recording formats
    c) Local weather forecasts
    d) Vendor agreements
16. A major concern of biopiracy is:
    a) Exploitation of indigenous knowledge without proper compensation
    b) Soil contamination
    c) Drought resistance
    d) Delayed germination
17. The Nagoya Protocol came into force in:
    a) 2000
    b) 2014
    c) 2010
    d) 1992
18. Which of the following is a valid reason for ethical opposition to GM animals?
    a) Concerns about animal welfare and genetic suffering
    b) Lower milk output
    c) Feed conversion ratio
    d) Tail length variation

19. A core ethical guideline in genetic research is to:
    a) Increase replication only
    b) Focus only on local varieties
    c) Minimize harm and maximize societal benefit
    d) Protect only elite germplasm
20. In biosafety research trials, a **Level II containment** is applied when:
    a) There is moderate risk to environment or human health
    b) There is no risk
    c) Animal testing is involved
    d) Secondary metabolites are needed
21. Informed consent from indigenous communities is essential before:
    a) Using their genetic resources or traditional practices
    b) Applying pesticides
    c) Building greenhouses
    d) Publishing a research article
22. Which international agreement supports rights of indigenous people in biodiversity?
    a) Kyoto Declaration
    b) UN Declaration on the Rights of Indigenous Peoples (UNDRIP)
    c) FAO Plant Treaty
    d) WTO TRIPS
23. A strong ethical policy in biotechnology must include:
    a) Only scientific risk evaluation
    b) Societal, cultural, and environmental dimensions
    c) Purely commercial interest
    d) Yield data reporting
24. A confined field trial must ensure that:
    a) The field is irrigated well
    b) The transgene does not spread to nearby fields
    c) All farmers support the research
    d) Government subsidy is granted

25. Which of the following is NOT a recognized biosafety level?
    a) BSL-1
    b) BSL-5
    c) BSL-2
    d) BSL-4
26. Traditional knowledge systems are often vulnerable to:
    a) Exploitation through patenting without consent or benefit sharing
    b) Natural disasters
    c) Germplasm erosion
    d) Cross-pollination
27. The scope of biosafety trials includes:
    a) Soil carbon quantification
    b) Testing ecological and health safety of transgenics before release
    c) Fertilizer application planning
    d) Seed dormancy management
28. One of the most ethically sensitive areas in biotechnology involves:
    a) Transpiration studies
    b) Human gene editing and heritable modifications
    c) Plant height increase
    d) Mineral biofortification
29. The principle of environmental justice in biotech ethics refers to:
    a) Fair distribution of environmental risks and benefits across populations
    b) Agricultural input subsidies
    c) Seasonal irrigation
    d) Regional productivity